D0301594

Human Factors
in Product Design

Human Factors in Product Design:
Current Practice and Future Trends

Edited by
W.S.GREEN and P.W.JORDAN

PARK CAMPUS
LEARNING CENTRE
C & G C.H.E. P.O. Box 220
The Park
Cheltenham GL50 2QF
Tel: (01242) 532721

TAYLOR & FRANCIS
ALERE FLAMMAM
Founded 1798

First published 1999 by Taylor & Francis,
11 Fetter Lane, London, EC4P 4EE

Simultaneously published in the USA and Canada by
Taylor & Francis Inc., 325 Chestnut Street, Philadelphia PA 19106

Taylor & Francis is an imprint of the Taylor & Francis Group

Copyright © Bill Green and Pat Jordan 1999

*All rights reserved. No part of this publication may be
reproduced, stored in a retrieval system, or transmitted in any
form or by any means, electronic, electrostatic, magnetic tape,
mechanical, photocopying, recording or otherwise, without the
prior permission of the copyright owner.*

Camera Ready Copy and text editing by
Stefan Flik (XD designers, Delft) and Ruth Green

British Library Cataloguing in Publication Data

A catalogue record for this book is available from the British Library.
ISBN 0-7484-0829-0 (cased)

Library of Congress Cataloguing-in-Publication Data
Jordan, Patrick W.
Human factors in product design: current practice and future
trends/Patrick W. Jordan and William S.Green.
p. cm.
Includes bibliographical references.
ISBN 0-7484-0829-0 (alk. paper)
1.   Design, Industrial. 2. Human engineering.
I. Green, William S.   II. Title.
TS171.4.J655 1999
745.2 — dc21
99-13215
CIP

Typeset in 10/11pt Times Roman
Printed and bound by T.J.International, Padstow, UK

Contents

Ergonomics, Usability and Product Development

CURRENT PRACTICE AND FUTURE TRENDS

WILLIAM S. GREEN and PATRICK W. JORDAN

DEDICATION

The editors and authors would like to dedicate this book to the memory of Prof. Dr. ir. Rudy den Buurman, who died during its production. Rudy was Chair of the Technical Products Group of the IEA, and chaired the meeting in Tampere where we first proposed the idea. His enthusiasm and support helped make it possible.

PART 1 — ERGONOMICS AND DESIGN

This book grew out of a meeting of the International Ergonomics Association Technical Products Group in Tampere, Finland. The meeting followed a highly successful symposium on usability in which several papers preceded a very lively, at times even heated, panel-plus-audience debate on the methods, theories and effects of user trialling and usability testing. The debate began in the usual fashion with questions directed at panel members which reflected the issues raised in their papers, but the real catalyst was a question from a Swedish designer, who said in very blunt terms that he was fed up with the usual academic waffle, and when were the academics going to provide really useful information to people like himself? His major point was that he was at least as far advanced as the panel members in terms of theoretical background and actual practice, and what he had come for was some forward thinking and help with directions for the future, not more of the same old stuff in different words. This brought an immediate and vigorous response from a number of people, and views were expressed which reflected a very wide spectrum of interests. Some speakers agreed, but most defended the papers, indicating that they had learned a lot from them. The chair took the view that if Dr.X was actually doing everything he said he was, then he was very advanced in terms of normal industrial practice and should perhaps be coming to inform rather than to be informed. His point was, however, well made, and considerable discussion revolved around the problems of translating laboratory research into useful and applicable data for designers.

A second thread of contention was the assertion from the chair that there was no acceptable theoretical base for research into user/product interaction. This outraged some of the psychologists present who proceeded to propose a number of theories, most well known, some rather arcane, as suitable bases for product interface design. As may be supposed, this discussion has no clear end, and it was euphemistically agreed that further work was necessary!

The large number of people attending the symposium was evidence of the current importance of the topic in both design and ergonomics circles. It was not always thus.

Ergonomics and design have had a long standing and curious relationship. Curious in the sense that what is evidently a symbiotic relationship has been polarised in ways both accidental and deliberate by both parties, sometimes to their mutual detriment, whilst some members of each set (including the editors and authors in this book) have made continuous efforts towards better communication.

The problem is partly due to history and education. 'Design' is primarily an art based discipline, and even industrial design has a history of having been taught predominantly in schools of art. 'Ergonomics', on the other hand, takes pride in its scientific provenance, in spite of (or perhaps because of) having a foot in both the physical and the social sciences camps. Much has been written on why designers don't use ergonomics data, and almost as much on why ergonomists don't present data in ways acceptable and usable by the design professions. Of course, designers do sometimes use ergonomics data, and ergonomists do sometimes present it in usable ways — the evidence is all around us of products and systems which work well enough, most of the time. This does not mean that we may ignore the extremities of the distribution curve where the really undesirable events take place, nor should we assume that, because something works, it works well. There are some substantial gaps between the *modus operandi* of the groups which makes for difficult dialogue and these may be pigeon-holed as the division between analytic and synthetic operations; science and art. It is precisely this division which is the grist of the usability mill.

Design may be seen as a creative endeavour in which originality, or even idiosyncracy, is valued and the errant genius lauded: emotional rather than rational; qualitative rather than quantitative; feeling rather than thinking. The industrial design engineers and product developers will react against that statement, but allowing for some deliberate extremism there is truth in it. It reflects a particular facet of humanity and an attractive one at that. Modern product development and industrial design engineering as a discipline, on the other hand, lays a claim to rationality which is based on the countable and measurable. These two meet most evidently where the user interacts with the product, and out of the meeting develops the functionality, efficiency, pleasure and emotional satisfaction or otherwise which that interaction brings, from purchase to discard. Ergonomics has provided a sound quantitative basis for many of the parameters of this interaction, and has brought design to the point where complete failures of safety or usability are, if not rare, then at least not often life-threatening. How much can be attributed to ergonomics in a formal sense and how much to designer self-reference is a matter of debate, and the absorption of formal ergonomics data into design processes is discussed later in this volume. What is clear is that purely technical competence is increasingly being taken for granted in modern product design, that the human aspect is becoming more prominent in product competition, and that ergonomics as a discipline has an opportunity as never before to establish its pre-eminence in the design portfolio.

The chapters in this book are drawn from the Tampere conference, but with many modifications and some additions, with the aim of telling a coherent story about the problems and possibilities of usability testing, human product usage and the future of ergonomics in product design and development.

The first part, 'Usability as a Commercial Issue', is included in this introduction and deals with one of the fundamental issues of usability — its value in commercial terms. Without this being established, most of the work we do in the field will be irrelevant, and the real test of a usability strategy is its effectiveness in a commercial environment. Usability is an important commercial issue and is increasingly being recognised as such. Because of this, human factors specialists have an unprecedented opportunity to influence the product creation process (Jordan, Introduction Part 2).

In part two, 'Usability in the Product Creation Process', some of the more theoretical and methodological issues are discussed. Usage centered design has considerable currency, but it is not reasonable to expect miracles from it. In order to be effective in the product creation process, human factors and usability specialists have to operate within particular philosophical, methodological and pragmatic frameworks. In the first chapter, some of the possible weaknesses of usability testing and user trialling are discussed (Green, Chapter 1) and the following chapters provide an elaboration of this and a defence to the criticism.

Part of working effectively with designers is getting the information across at the correct time in the design process and in the correct manner. Some examples are given of how this can be achieved (Porter, Chapter 2).

Usability trialling is an essential tool for the human factors specialist. Some of the most commonly used methods, with examples of the context in which they can most usefully be applied, are discussed by Popovic (Chapter 3). Research into user activities has some particular methodological issues which are discussed, first by Kanis (Chapter 4) in the context of a carefully developed conceptual model, and also in detail by Vermeeren (Chapter 5) who uses a particular experiment to explore control versus reality in usability trials.

'Smart' products are emerging as a new category of consumer electronics product. Some reflections on the emergence of smart products and some predictions about their future development are provided (Bonner, Chapter 6).

Iterative prototyping and evaluation are beneficial to the creation of usable smart products. Some prototyping techniques that are particularly suited to this type of product are described along with their advantages and disadvantages (Sade, Chapter 7).

The safety of the user must always be uppermost in the mind of the product designer. Norris (Chapter 8) raises a number of major issues, with advice as to how the designer or human factors specialist can address these.

Part three, 'Case Studies' is concerned with examples and case studies drawn from various contexts.

Conducting trials on early concepts can move the design process in the right direction (Hall, Chapter 9) and the dangers of not doing so are illustrated by Green and Klein (Chapter 10).

Often, industry-based human factors specialists have to work under tight financial and temporal constraints. Examples of how to get the best out of an evaluation conducted under those circumstances are provided by Thomas (Chapter 11) and Kahmann (Chapter 12).

Software based prototyping can be a very useful tool in the evaluation process. However, its effectiveness may be questionable. An example of how such tools can be used and the tools themselves evaluated is provided by de Vries (Chapter 13), whilst the study by Rooden (Chapter 14) looks in a detailed way at the value of different early product representations in user trials.

Finally, even though 'best practice' has been followed, a product can sometimes fail for unforeseen reasons. The case study by Trathen *et al.* (Chapter 15) provides valuable lessons for those involved in product design and evaluation.

Part four, 'Inclusive Design — Design for All', deals with the social and demographic changes occurring in the developed nations. The aging population brings with it a number of challenges and opportunities for design. These are outlined and strategies for addressing these challenges and opportunities are proposed (Coleman, Chapter 16). Disability is not just about old age, and the moral imperatives of design for all are discussed by Jordan (Chapter 17). A case study by Simpson (Chapter 18) shows a process of design for people with sensory deficit, in this case reader systems for the blind. Consumers' representatives are also concerned with the issue of inclusive design. Here a representative of the Research Institute for Consumer Affairs discusses the methodology that they use to evaluate products for inclusive design (Etchell, Chapter 19). This methodology addresses mainly the physical aspects of human-product interaction and in the following chapter Freudenthal (Chapter 20) is concerned with the sensory and cognitive problems which arise, and suggests some design guidelines which are appropriate for a wide range of product users.

Part five, 'Beyond Usability — Pleasure with Products', is concerned with a view of the future. Fitting the product to the human is about more than just usability. Human factors should look beyond usability to the factors that make a product a positive pleasure to use. A framework for considering pleasure with products is outlined (Jordan, Chapter 21).

An analysis of the product characteristics that influenced users' perceptions of quality, indicate that psycho-social factors extending beyond usability are important. A case study is used to illustrate this (Taylor, Chapter 22).

The Japanese technique of Kansei Engineering is an example of pioneering work in the area of analysing links between design and pleasure. This is described and some examples of its application are given (Jordan, Chapter 23). Finally, a method for assessing the tactile and sensory aspects of a product is outlined and its application described (Bonapace, Chapter 24).

The book concludes with a discussion of the issues raised and addressed, and some assertions about suitable future directions for research and practice.

INTRODUCTION: PART 2 — USABILITY AS A COMMERCIAL ISSUE

Summary

Human factors input to the design process is seen by manufacturers as giving their products a competitive edge in the marketplace. Customers have come to expect and demand that products are easy to use. However, if human factors is to maintain its competitive relevance, the discipline must make a contribution that goes beyond usability to also address the emotional and hedonic aspects of product usage. This means looking at users holistically — as human beings rather than simply as physical and cognitive processors.

Introduction

As a discipline, human factors is now in a position of unprecedented strength. Reflections of this are, for example, the ever growing human factors literature, the number of national and international conferences that cover human factors issues and the extensive research on human factors topics that is carried out in academia. However, the most significant reflection of how seriously human factors issues are now being taken is, perhaps, the increasing number of human factors specialists employed by industry.

Human factors as a competitive advantage

Why has human factors come be in such a strong position? Clearly, if industry is prepared to spend money addressing human factors issues then companies must perceive that they are getting some commercial benefit in return.

Manufacturers are constantly seeking ways of gaining an advantage over the competition. In this advanced technological age, the ability to make gains in 'traditional' ways — such as technological reliability or manufacturing costs — are being continually eroded. As a result, design has become an issue of increasing commercial importance.

Good human factors is seen as a central tenet of good design. Manufacturers are advertising products as 'easy to use' or 'ergonomically designed' and consumers are becoming more sophisticated in terms of what they expect and demand of a product in terms of its human factors qualities.

Whilst people may once have accepted difficulties in usage as the price to pay for 'technical wizardry' this attitude has changed. People will be disappointed with products that are not well designed from the users' point of view. Of course these products will also end up disappointing those who manufacture them as, sooner or later, people will stop buying them.

Usability — human factors' contribution to product development

Human factors input is thus seen as giving products a competitive advantage. But what form does this input take?

Traditionally, human factors input to product development has tended to be focused on usability issues. The International Standards Organisation (ISO) defines usability as being:

> *"... the effectiveness, efficiency and satisfaction with which specified users achieve specified goals in particular environments."* (ISO DIS 9241-11)

Effectiveness refers to the extent to which a goal, or task, is achieved, efficiency, to the amount of effort required to accomplish a goal. Satisfaction, meanwhile, refers to the level of comfort that the user feels when using a product and to how acceptable the product is to users as a means of achieving their goal.

Human factors as a profession has been very successful in terms of quantitative definitions of these concepts. For example, effectiveness might be defined in terms of the percentage of attempted tasks that a user is able to complete with a product, efficiency in terms of the time taken to complete the tasks or the number of errors made along the way. A number of questionnaire based attitude measurement tools have been developed for the quantification of satisfaction. These include the System Usability Scale (SUS) (Brooke 1996) and the Software Usability Measurement Inventory (Kirakowski 1996).

Similarly, the profession has made a great deal of progress in terms of identifying the properties of a product that make it usable (see, for example, Ravden and Johnson 1989) and in developing methods for the evaluation of usability.

The progress in these areas has been central to the success of human factors in becoming established as a fundamental part of the product creation process within many commercial organisations. In particular, the quantification of usability takes the issues into a concrete domain that can be understood by others involved in the product creation process, such as engineers and product managers. Quantitative usability targets can also be written into the product specification in a similar way to technical specifications. This affords human factors specialists far more influence in the product creation process than they would have if they could only talk in abstract or qualitative terms about 'quality of use', 'ease of use', 'ergonomic design', etc.

The next step — a holistic approach

Usability issues are now taken seriously in the product creation process within many organisations. This is, of course, encouraging. However, if human factors specialists concern themselves only with usability, they are still falling short of assuring the optimisation of a person's experience with a product.

Usability based approaches tend to look at products as tools with which users achieve tasks effectively, efficiently and within certain levels of comfort and acceptability (ISO DIS 9241-11). In reality, though, products have a far wider role in people's lives. They are not merely tools but are 'living objects' with which people have relationships. The next challenge for human factors is to move from looking merely at *users, products and tasks*, to looking, more holistically, at *people, products and relationships*.

People are complex, rational and emotional beings. Usability engineering approaches often tend to look at people as merely being part of a three-part system — product, user, environment of use. This type of approach is in danger of dehumanising people — looking at them as physical and cognitive processors. Issues such as people's emotions, values, hopes and fears are often ignored. This is to ignore the very essence of what makes us human.

The new challenge for human factors is to look not only *at* but also *beyond* usability. To look at the other factors that can affect the quality of a relationship between person and product. People/product relationships are also likely to be affected by issues such as the look and feel of a product, the fit of the technical specification and performance with user expectation, and the emotions and values that are associated with a product. Human factors must be wary of concentrating on usability to the exclusion of these other vital issues. In practice this will mean looking beyond the current tenets of ergonomic design (e.g. design for consistency, compatibility, feedback) and the traditional measures by which usability is quantified (e.g. task success, time on task, error rate). It is important to also look at the emotional and hedonic benefits that products can offer people and to investigate what the implications of these will be for design and evaluation. So far there has been little human factors work into these wider issues of people/product relationships (Dandavante, Sanders and Stuart 1996, Rijken and Mulder 1996, Jordan and Servaes 1995, and Jordan 1996 are rare examples).

These new challenges for human factors are expanded upon in the last section of this book and techniques for addressing these issues during product development are outlined (e.g. Jordan, Chapters 21 and 23; Taylor, Chapter 22; Bonapace, Chapter 24).

Conclusion

Human factors input into the products development process can give products a competitive advantage in the marketplace. Users expect and demand usability and are increasingly unwilling to settle for less. This is both a tribute to the human factors profession and a reflection on the importance of good design as a competitive issue. Now human factors can move a stage further, taking a holistic approach to the analysis of people/product relationships. This will support the creation of products that are not only useful and usable, but which also bring emotional and hedonic benefits. Such products will delight customers and keep manufacturers ahead of the competition.

REFERENCES

Brooke, J., 1996. SUS — A quick and dirty usability scale. In *Usability Evaluation in Industry*, P.W. Jordan *et al.* (Eds), Taylor & Francis, London.

Dandavante, U., Sanders, EB-N and Stuart, S., 1996. Emotions matter: user empathy in the product development process. In *Proceedings of the Human Factors and*

Ergonomics Society, 40th Annual Meeting — 1996, Human Factors and Ergonomics Society, Santa Monica, pp. 415-418.

ISO DIS 9241-11, Ergonomic requirements for office work with visual display terminals (VDTs): - Part 11: *Guidance on Usability*.

Jordan, P.W., 1996. Displeasure and how to avoid it. In *Contemporary Ergonomics 1996*, S.A. Robertson (Ed.), Taylor & Francis, London.

Jordan, P.W. and Servaes, M., 1995. Pleasure in product use: beyond usability. In *Contemporary Ergonomics 1995*, S.A. Robertson (Ed.), Taylor & Francis, London.

Kirakowski, J., 1996, Software usability measurement inventory; background and usage. In *Usability Evaluation in Industry*, P.W. Jordan *et al.* (Eds), Taylor & Francis, London.

Ravden, S.J. and Johnson, G.I., 1989. *Evaluating Usability of Human Computer Interfaces: A Practical Method*. Ellis Horwood, Chichester.

Rijken, D. and Mulder, B., 1996. Information ecologies, experience and ergonomics. In *Usability Evaluation in Industry*, P.W. Jordan *et al.* (Eds), Taylor & Francis, London.

Bill Green and Pat Jordan
Delft and Groningen 1998.

CHAPTER ONE

Essential Conditions for the Acceptance of User Trialling as a Design Tool

W.S. GREEN

Faculty of Industrial Design Engineering, Delft University of Technology,

Jaffalaan 9 2628 BX Delft, The Netherlands.

This paper was originally intended as a precursor to the panel discussion on usability testing in the usability symposium at the IEA conference, Tampere, Finland, in 1997 and as a stimulus for debate. It assumes a 'devil's advocate' position and questions the value of user trialling from three positions: a general level which may loosely be termed philosophical, the methodological position, and the pragmatic dilemma. It concludes with a discussion of some of the requirements of successful usability testing.

1.1   INTRODUCTION

In contrast to the previous millennia of small incremental adjustments with occasional jumps, the recent history of man product interaction has been characterised by discontinuous change. The industrial revolution was the first, the two world wars the second, and the information revolution the third. The potential for failure in increasingly complex man machine systems began to be formalised during the second of these stages, and various models were proposed for safe system design. In the absence of alternatives, the inexplicability of disasters was frequently labelled as 'human error', and the apparent capacity of machines to 'fight back' was documented by John Stapp in 1949 in response to Edward Murphy's irritable pronouncement that "if anything can go wrong, it will" (Block 1990). This has become universally and humorously known as Murphy's Law, and has been extended by popular assent (Murphy's second law) to say that it will happen at the time of maximum effect (or, to quote a popular example, the chance of the bread landing jammy side down is directly proportional to the cost of the carpet). I have proposed (Green 1997) that there should be a third law of Murphy, which says that "Laws 1 and 2 are not reliable" because there is an implicit assumption of predictability in the original Murphy's Law(s). Predictability is at the centre of the problem, and the early history of ergonomics has predisposed us to believe that human actions are indeed usefully predictable. In the early (military) populations studied, this was true to some extent. When a population can be controlled and trained and its equipment and procedures specified to such a high degree, then it is easy to see why the biggest variables could be thought of as human characteristics, rather than as behavioural, variables. However, when a domestic situation is considered, it is clear that not only is the anthropometric variability extreme, but also that the level of training and control is very low indeed, and human capacities, even if documented, are no longer reliable indicators of how people will interact with products. Predictability in any situation is to some extent a function of resolution, but certainly may not be taken for granted. When natural systems which include humans and machines are juxtaposed, strange and sometimes undesirable things can happen. Knowledge of human characteristics has been assiduously applied to the design of products and things still go wrong, *because allowing the possibility of use does not control the actual usage.* Given the lack of a theoretical framework of user/product interaction, usability testing is being flagged as the only available technique for minimising user/product mismatch. If it is indeed the only technique, then it is as well to be clear about its limitations.

1.2   IS THERE A GENERAL PROBLEM?

The term we apply to the more serious of the undesirable unpredictable events deriving from man-product-environment interaction is 'accident'. An accident may be described as an 'event without apparent cause'(Oxf. Dict.) In other words, there is no identifiable causal relationship between what happened and the factors surrounding the event. If this is so, then there is no chance of ever exercising general control over 'accidents'; this is supported by Weegels (1996). However, the term is not applicable to all of the myriad problems arising from product use, ranging from minor inconvenience, irritation and inefficiency, to physical harm. The implication is that, for many of these occurrences, there *is* a causal link — that the outcome is predictable at least in probabilistic terms — and this assumption underpins the research fields of usability testing and user trialling. Indeed, it is a very significant primary assumption in any design endeavour. It can be argued, however, that not only are the links difficult to observe, but the observable links are of limited use. Certainly, it is useful in any given case to determine what are the effects of real user behaviour with a product, but as has been shown (Kanis 1993: 95) the range of human diversity is large, and the very nature of usability testing usually prohibits anything more than a very small sample of the possibilities. When something goes wrong, we speak of so-called deviation, but what is deviant, and from what does it deviate? Most things work well enough most of the time, and when an undesirable event takes place it is convenient to accept that the source of deviation lies within the human actions. However, as Weegels and Kanis (1998) have observed, humans may be tired, distracted, angry, in a hurry and so on, and these are not criminal or even unusual states. Furthermore, the local environment or context may be highly specific. The chances of encountering 'deviant' behaviour in a small sample is small (Kanis and Vermeeren 1996), and what we require of a domestic product design is efficient, safe and even joyful operation for both likely and, perhaps, highly unlikely usage. In other words, we require a synthesis, in the design, of those cues and attributes, which precipitate desirable usage regardless of context or human proclivities. Kirlik (1995) quotes Braitenberg (1984) as suggesting that human capacities for synthesis far exceed capacities for analysis. There is an inference of an almost mystical nature in this, suggesting support for the 'designer as black box' (see Jones 1970) in which unknowable connections are made and out of which comes a capacity to get things right. This was the acceptable *modus operandi* of designers for many years and for many it still is, and the simple reason for its robustness is that, often enough, it appears to be true!   Should we leave designers to get on with it? Clearly, the 'magic' does not seem to work with so-called 'smart' or indeed unfamiliar products, but are the alternative strategies significantly better?

Millennia of development and evolution have provided us with a base from which we inevitably operate. It is essentially a physical base, with action following perception as directly as possible. A product interface, which requires reasoning, will always provide more difficulties for us than one in which action and response have a clearly observable link. The proposal has frequently been made anecdotally, that all we have to do is wait for the generation of children versed in the way of electronic gadgetry to grow into consumers and all the problems will be solved, but this seems to be at the very least a little naive and may be a gross misunderstanding of the relationship between perception and cognition. The evidence we have points to the evolution of new and more complex products as always being faster than our ability to adapt and thus the maintenance of a permanent operational chasm for which we seek bridges. Usability testing and user trialling is certainly a potential bridge, but is it the best? Is the outcome of usability testing better than that which informed designers already achieve?

There are three terms, which might usefully be applied to the field of study, dealt with in this paper: usage, usability, and functionality. The accusation of hair-splitting can be predicted, but when detailed consideration is given to the research required in the area, useful distinctions can be drawn. There is a hierarchy of resolution in the terms. Consideration of usage involves the mapping of perception/cognition and use actions via conceptual models, physical models and prototypes or actual products. Usability is the impact of the amalgam of human characteristics and mental models on product performance. Functionality may be seen as extending these to including impact on quality of life — whether a product will be bought, or used at all, and involving notions such as joy in use, effective performance, perceived stigmatisation, actual or perceived safety. Thus functionality can claim precedence in the order, and there are two important terms when discussing product functionality, which have already appeared in the text without amplification. These are 'predictability' and 'resolution', and they are closely linked. Predictability is at the root of all theoretical effort: if a theory has no predictive value it is essentially an historical concept, interesting and perhaps useful as a *post facto* explanation of events or phenomena, but of no great consequence to, for example, a busy product development engineer. In design terms (maybe in any terms), predictability is a function of resolution. It has an unfortunate tendency to diminish with higher resolutions, which forces mass production design effort into the low-resolution domain. With the current move towards customisation of products, this presents a problem for manufacturers and product development engineers. A non-handicapped person wishing to leave a room will behave, under normal unstressed circumstances, in a highly predictable manner. Given a door with a lever operated latch mechanism, they will move to the door, depress the latch lever, open the door and leave. At this resolution a high level of predictability is evident. However, the amount of force they will apply to the lever is much less predictable, and the precise location and distribution of their fingers on the lever is even less predictable, and so on. In this case it is relatively easy to decide that the design significance of the higher resolutions is low. It doesn't matter greatly how hard they press the lever, provided it is hard enough, and where the fingers are on the lever may happily be ignored. Our entire product environment works on these design assumptions which have been labeled 'population stereotypes' — the expectation of 'normality' based, usually, on the designer's definition of the term. Donald Norman (Norman 1988) has explored this issue in some detail, but there is still no predictive methodology for dealing with levels of resolution, which are higher than previously applied and experienced. In any given case, what may be called the threshold of resolution for successful operation may only be determined by experience, and a new interface is, by definition, new.

1.3   THE METHODOLOGICAL PROBLEM

That there are such problems is not at issue. I choose to describe them in three ways:

1.   Cleanliness in the test procedures

We have no clear answers to the questions of how to map product use. For example, what the relationships are between model and reality; how well do so-called 'expert' techniques such as cognitive walkthrough and heuristic assessment relate to three dimensional products (see Green, Kanis and Vermeeren 1997), and how well do laboratory studies reflect actual in service use actions. Experiments with drawings, primitive models, developed models and real products have done little more than produce further uncertainty (Rooden and Green 1997; Rooden 1998). In some cases the performance of *ab initio* users with primitive models was 'better' than with highly developed (and thus supposedly more informative) models. Techniques such as Thinking Aloud (TA) as proposed by Ericsson and Simon (1993) have value in highly structured circumstances, but may be limited when applied to product use involving a combination of perception/action and problem solving engagement. TA is only really applicable to 'knowledge based' (Rasmussen 1986) behaviour, and it is unclear how such behaviour may be determined in product use in natural environments (see Rooden and also Popovic in this volume). In any case, Rooden (op. cit.) found that his subjects became silent when faced with difficulties in operating unfamiliar products, and that their verbalisations were in fact action reports, rather than illuminations of thought processes. Observation in natural environments with unobtrusive cameras provides the best source of uncompromised data, but presents an analysis problem unless linked with other supporting or confirmatory procedures (Weegels 1997) and is only really possible with actual products or at least working prototypes, which limits its use as a design tool early in the product development process. Later in this volume, Heimrich Kanis explores these issues in greater detail.

2.   Generalisability

There is a strong case for saying that it is better to trial than not, in spite of the possibly poor 'signal to noise' ratio. There is professional acceptance of the value of even primitive trials (Green and Polanen-Petel 1996). However, this value may only be confidently applied specifically. The transferability of data gained in one set of circumstances to another set is questionable. The cues which prompt particular actions are incompletely or poorly understood, and we have no firm way of knowing what may be transferred and what may not. Thus user trialling becomes a simple one-off technique which may identify localised problems. Repeatability is questionable even if possible, and we are left with some hazy notions of what might have an effect on user actions if all the circumstances are more or less as expected. Doesn't sound very convincing, does it?

The question of generalisability raises the issue of a theoretical basis for human/product interaction. The established theories of e.g. Gibson and Rasmussen are robust and valuable for observing and explaining what happened and perhaps even why it happened, but are not very useful as predictors of product interaction. It is certainly not very valuable for a designer to be told "design for use at Rasmussen's first level". The first response is likely to be "What?" and the second a dismissal of the issue as 'something we do anyway'. Recent works by researchers such as Kirlik and Vicente have espoused the principles and highlighted the research problems of an ecological approach to interface research. A little diagram by Kym Vicente (Vicente 1997) gives a neat (and rather gloomy) appraisal of the research situation by showing a see-saw curve with researchabilty and relevance at opposing ends. The more you have of one, the less you have of the other! The issue is also explored later in this volume by Kanis. This indicates the dilemma of researching what people actually do, and by inference the difficulty of ever formulating a predictive theory for interface design for products. It is no coincidence that extensive discussion with colleagues who are working in the area yields the same conclusion: that the best way forward is to concentrate on simply 'doing it', the assumption (or rather, hope) being that a 'body of findings' accruing from a bottom up approach will eventually coalesce into a workable set of principles or theories. The possibility of a 'grand unifying theory' which has genuine predictive value is very small.

3.   Design

When all of the tests have been conducted and the recommendations made, they are given to a designer, or design team, who will interpret them and incorporate them into the design, making such adjustments and accommodations (to aesthetic values, production processes and so on) that it is the designer's brief to do. How these adjustments may change the (uncertain) values of the trial results is unclear. They may enhance, they may compromise. We do not know. In much of the writing on usability testing there is an ignorance of the actual processes undertaken by product designers and developers — of the imperatives that many designers feel as a result of their education, training, work situation and their association with their professional colleagues. These imperatives are frequently masked by formalised design process methodologies, which are comforting for everybody concerned but which may not illuminate the real effects of other pressures. In the product development process there are a series of points of judgement when a person or persons must do what they feel is best, given the fluidity of the information with which they are presented. That person is probably the designer(s). Isn't that what designers have always done? Haven't we just arrived back at the circle entry point? Obviously, the argument goes that the better-informed designers are, the more likely they are to make good ergonomic decisions, and it is a seductive argument. However, if the history of designers' use of standard ergonomic data is considered, then we are in for a long wait before such new data are satisfactorily absorbed into the design lexicon.

1.4   THE PRAGMATIC PROBLEM

This is presently one of the most intractable of all of the problems facing user-centred designers and usability experts. It is comforting to talk to like minded people in a medium such as this, and to agree that this or that is valuable, this or that will help the user, this or that is clearly the case. Comforting, but not very useful. It is the time-honoured practice of preaching to the converted. The people who are reading this are probably not those to whom we should address our efforts.

Simply put, we don't have a lot of detailed and dependable information and even the information which we have is not penetrating the problem. The reasons are complex. The real design process is an amalgam of compromises, for which many idealised models have been proposed, but which may reliably conform to none of them. There are many imperatives, and in the event of usability being one of them (and this is not always the case) it must take its place in the hierarchy — a hierarchy in which marketing, aesthetics and production tend to be dominant. Thus we have a process of questionable value which may or may not be known, or if known not applied, or if applied ineffectively.

1.5 DISCUSSION

There are thus certain necessary conditions for the application of user centred studies.

1. *At a general level* we must be convinced that there is a possibility of change and a need for change and that there is an understanding of the direction of movement. The designer as 'black box synthesiser' is a model of operation not without its attractions. Evaluated as a ratio of product use acts against accidents, or even goal failures, it can be seen to work quite well most of the time. Is there a convincing case for changing it? There is considerable agreement that the Rasmussen 'entry' level of *see....do* is the goal of product interaction. An intuitive approach with minimal cognitive involvement leads to the most direct and effective product use, and this is what many if not most designers strive to achieve. (Minimum cognitive engagement is clearly not always desirable, as for example in the design of a recreational puzzle, but in terms of domestic product design it may be considered a useful aim.) How it may be achieved is another matter.

2. *Methodological.* To be genuinely powerful, research must be directed toward the generalisability of data. If every design has to go through an exploratory process which investigates its every aspect via user trialling, we may expect commercial resistance to a degree which renders the process inoperable. Even within these constraints, there are problems of method which are scrutinised in the chapters which follow, but sooner or later we have to expect that some general rules will emerge at a level of abstraction which is sufficiently low to be genuinely useful for busy product designers.

3. *Pragmatic.* The results of user trials must be directly applicable by designers. They must, seamlessly and with zero cost, be amalgamated into the design process, or they must be seen by manufacturers to convey a significant competitive advantage.

There is an interesting tendency in these conditions. The designer appears to be central to the usability generation, evaluation and implementation process. A general theory is unlikely. Much more likely is a growing compendium of case studies which can not in themselves produce powerful generalisations, but which may allow a sophisticated design synthesis to occur when the designer(s) has access to the data. Design synthesis works best when the data are internalised — in other words, when the designer has experienced the data generation. The application of the data in design terms is less likely to be a corrupt process when the designer is using self-generated data, and the efficiency with which the data is incorporated into the design process is enhanced immeasurably if the designer is intimately involved. Seeing clients struggling with a design which is a designer's pride and joy can be a seminal experience for the designer involved. Commercially, the prognosis is much more promising. As Pat Jordan has pointed out 'user friendliness' is a significant marketing factor, certainly in the computer field, and as more and more domestic products acquire digital interfaces, we may expect the demand for such user friendliness to spread. Perhaps not so much a case of competitive advantage if present, but competitive death if absent. Whether our current techniques of user trialling can provide it is the key question.

1.6 REFERENCES

Block, A., 1990. *Murphy's Law Complete.*
Braitenberg, V., 1984. *Vehicles: Experiments in Synthetic Psychology,* MIT Press, Cambridge, MA.
Ericsson, K.A. and Simon, H.A., 1993. *Protocol Analysis,* MIT Press, Cambridge, MA.
Green, W.S., 1997. *Predictability of Product Use, or Murphy's 3rd Law,* Faculty of Industrial Design Engineering, Delft University of Technology.

Green, W.S. and Polanen-Petel, F.V., 1996. A methodology for user trialling. In *Enhancing Human Performance,* Proceedings of the 32nd Annual Conference of the Ergonomics Society of Australia.

Green, W.S., Kanis, H. and Vermeeren, A.P.O.S., 1997. Tuning the design of everyday products to cognitive and physical activities of users. In *Contemporary Ergonomics,* S.A. Robertson (Ed.), Taylor & Francis, London, pp 175-180.

Jones, J.C., 1970. (+ reprints) *Design Methods: Seeds of Human Futures,* John Wiley & Sons, Chichester.

Kanis, H., 1993. Operation of controls on consumer products, *Human Factors,* **35**, pp. 305-324.

Kanis, H. and Vermeeren, A.P.O.S., 1996. Teaching user involved design in the Delft curriculum. In *Contemporary Ergonomics,* S.A. Robertson (Ed.), Taylor & Francis, London, pp. 98-103.

Kirlik, A., 1995. Requirements for psychological models to support design. In *Global Perspectives on the Ecology of Man-Machine Systems,* John Flach *et al.* (Eds), Erlbaum Associates, NJ.

Norman, D.A., 1988. *The Psychology of Everyday Things.* Basic Books, N.Y.

Rasmussen, J., 1986. *Information Processing and Human Machine Interaction: An Approach to Cognitive Engineering,* Amsterdam.

Rooden, M.J. and Green, W.S., 1997. Anticipating future usage of everyday products by design models. In *Proceedings of the 13th Triennial Congress of the International Ergonomics Association,* **2**, W.S. Green and M.J. Rooden (Eds), Tampere, Finland, pp. 168-170.

Rooden, M.J., 1998. Thinking about thinking aloud. In *Contemporary Ergonomics,* M. Hanson (Ed.), Taylor & Francis, London, pp. 328-332.

Valery, P., 1962. Unpredictability. In *History and Politics,* transl. Folliot and Mathews, Pantheon Books, N.Y.

Vicente, K.J., 1997. Heeding the legacy of meister Brunswick and Gibson: toward a broader view of human factors research. In *Human Factors,* **39**, pp. 323-328.

Weegels, M.F., 1996. *Accidents Involving Consumer Products.* Faculty of Industrial Design Engineering, Delft University of Technology.

Weegels, M.F., 1997. *Exposure to Chemicals in Consumer Product Use.* Faculty of Industrial Design Engineering, Delft University of Technology.

Weegels, M.F. and Kanis, H., 1998. Misconceptions of everyday accidents. In *Ergonomics in Design,* **6**, 4, pp. 11-17.

CHAPTER TWO

Designing for Usability; Input of Ergonomics Information at an Appropriate Point, and Appropriate Form, in the Design Process

SAMANTHA PORTER

Visual and Information Design Research Centre, Coventry School of Art

and Design, Coventry University, Priory Street, Coventry CV1 5FB, UK

and

J. MARK PORTER

Department of Design and Technology, Loughborough University,

Loughborough LE11 3TU, UK

2.1   INTRODUCTION

The design of a successful product is dependent upon incorporating information from a number of different knowledge domains; it must perform at a functional level, at an aesthetic level and must also be manufacturable by the right process for the right price. The necessary information comes from a variety of disciplines including design, ergonomics and engineering. In the case of ergonomics, it would seem obvious that products are made to be used and should function properly, reliably and safely. However, products in the market place are not always designed with the user populations' needs, abilities, skills or preferences in mind; a user population which is likely to be diverse including people of all ages and functional abilities. At best a product/user mismatch may cause only inconvenience or discomfort (Porter, Porter and Lee 1992), at worst injury or fatality. Department of Trade, Home Accident Safety System (HASS) data (published yearly) shows that inadequate domestic product design does indeed cause serious injury and even fatality. The rather serious implication of this is that designers may not have an understanding of both the physical and the psychological human characteristics of the population for whom they are designing. There is evidence in the literature to support this notion. Burns and Vicente (1994), Hasdogen (1995) and Pheasant (1995) present many instances, in the design field, of ergonomics information being used inappropriately by designers, if it is used at all. In this chapter we aim to provide an example of good incorporation of ergonomics principles and data into the design process, discuss the reasons why poor practice occurs and make suggestions for improving practice.

2.2   THE LIGHTWEIGHT SPORTS CAR

A car has recently been designed at Coventry School of Art and Design where the traditional design practice of designing a car from the 'outside-in' (from exterior style through engineering to occupant, Tovey 1992) was ignored in favour of designing the car from the 'inside-out' using both old and new technologies and methods (Porter and Saunders 1996). The potential occupants were the starting points for all aspects of design and one of the aims of the design team was to successfully incorporate ergonomics into the design leading to successful design for usability.

The brief specified a sports car; weighing less than 500kg and costing less than £10,000 to be designed and built within 20 months. The weight, price and time constraints meant that concurrent design, engineering and ergonomics was essential. As a completely new vehicle there were none of the conventional vehicle design constraints (e.g. number and configuration of wheels and passengers) or a fixed mechanical package beyond a notional engine. It was, therefore, possible to define a potential user population and design for them before having to consider any engineering hard points.

SAMMIE (a 3D human modelling computer aided ergonomics design system, Porter *et al.* (1996) was used to define space envelopes and seat travel for the chosen user population. The users were defined by the design team as ranging from a small Japanese male (5JM), using 5th percentile values where appropriate (which is equivalent to an approximately 25th percentile European female in stature), to a large American male (99AM) using 99th percentile values where appropriate. The necessary 'minimal' design of the sports car means it will be competing with the likes of the Caterham 7 and the population was chosen to reflect that sports car buying population. They are predominantly purchased and driven by males, hence the relatively large 'smallest' user and the extremely large 'largest' user (Figure 2.1 shows typical output). The man models were set in appropriate postures for such a car, whilst conforming to the recent joint angle comfort recommendations of Porter and Gyi (1998).

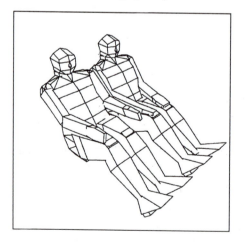

Figure 2.1 Example of SAMMIE output, used as underlays for sketches, a basis for design in Alias and to create 'footprints'.

The generated space envelopes were plotted at 10th scale in front, side and plan elevations and then used (in the traditional fashion) on paper in conjunction with basic engineering information (engine, drive train and wheels) in different configurations covering rear, front and mid-engined, three and four wheels in order to assess vehicle 'footprints' in plan view (see Figure 2.2, for example) and to choose one theme on which to develop the design. A conventional side-by-side, rear engined package was subsequently chosen, a decision based on several factors including the aesthetic potential, engineering feasibility, weight and cost.

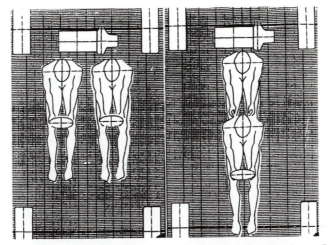

Figure 2.2 Typical 'footprint' used in the evalution of people and engineering configurations.

The designers subsequently developed an aesthetic theme based upon the brief, the package and their initial concept sketches (these are the first sketches designers produce and their intention is to convey the emotion and feel of the vehicle). It was also possible to import the SAMMIE 'men' into Alias, a 3D design tool which allows the modelling of complex curved surfaces and allows easy manipulation of form, colour and finish, Figure 2.3, and which also allows concurrent 3D CAD modelling alongside the conventional 2D aesthetic theme sketching programme.

The car was designed in Alias, and on paper, to 'look good' and to fit around the SAMMIE human models. A full size buck was concurrently built with seats, pedals, chassis tubes and nose cone for early ergonomics fit and visibility evaluations as well as aesthetic judgements. The provision of ergonomics constraints early in the project allowed an aesthetic theme to be developed which the designers were happy with while accommodating an appropriate population. The method of providing 10th scale space envelopes provided the perfect means of conveying the ergonomics data to the designers in a form they not only understood but on to which they could directly sketch the aesthetic themes. The importing of the SAMMIE mannequins into Alias was essential to the development of the design using a computer-aided styling tool.

The nature of the vehicle, combined with its weight and cost constraints, meant that a glass fibre, fixed back angle, bucket style seat was the most sensible solution. The design was to be in-house and further work was carried out using SAMMIE to establish the required seat adjustment. The studies resulted in the proposal of a trunk-thigh angle of 110 degrees (within the acceptable limits as recommended by Porter and Gyi 1998) and a forward adjustment of 200mm. The recommendation also included the raising of the seat combined with a tilting forward action to make the backrest more upright and to lower the seat front as forward movement occurred. This has the effect of improving forward sight lines for small members of the specified population as well as improving pedal reach for the smaller driver. The seat adjustment was combined with the pedal size and spacing as specified by Henry Dreyfuss Associates (1993). Some compromises were made as a consequence of engineering and aesthetic requirements. The seat travel was reduced to 165 mm to acommodate mechanical packaging requirements and the pedals were moved inboard (to accommodate the front suspension) by 130mm from the ideal. The pedal spacing is particularly wide for a sports car with the accelerator pedal positioned to the right of the seat centre line. It was possible to confirm 'showroom' acceptability of this package using the seating buck and a test population which included the size extremes of the population. The seating buck was also used to establish the acceptability of the exterior and interior visibility and an acceptable fixed steering wheel position.

18

Figure 2.3 A 3D Alias model of the vehicle.

At the next stage the seat design was developed in clay based on basic minimum (e.g. seat length) and maximum (seat width and seat back height) anthropometric dimensions. The designers were reluctant to include a head restraint as a part of the design, claiming that the additional 15-20cms on the seat would spoil the 'look' of the vehicle. Interestingly, the development of the head restraint in clay in the chassis received immediate approval, the height giving balance to a low and wide vehicle. The seat shape was developed with subjective feedback from both ends of the target population resulting in a seat which has 'showroom' acceptable comfort for both small and large users. Minimal cushioning for the seat was also designed, its positioning and shape based upon the pressure distribution plots of Porter and Gyi (1998).

This project demonstrates that if ergonomics input occurs at the appropriate time and fashion in the design process then the design can perform at an engineering level and remain aesthetically uncompromised. It also shows that using the appropriate type of ergonomics tool, and output, allows designers to use the information easily and readily. The SAMMIE system has been used in a similar way in many other projects (see Porter, Case and Freer 1996 for some examples), where the highly visual nature of the evaluation was particularly appropriate given an additional barrier to communication experienced in multi-national project teams — that of language itself. The lightweight sports car project was particularly successful because of the inclusion in the team of a permanent ergonomist who had an active role in the design process; providing the appropriate information, in an appropriate form at an appropriate time. The work of Haslegrave and Holmes (1994) also identifies the inclusion of an ergonomist as a permanent team member as a major contributing factor to successful ergonomics design. However, for large numbers of small design consultancies this is not an economic reality and it is designers who are responsible for the ergonomics input to a design. Examination of the work of both ergonomists and designers raises a number of issues.

2.3 PROBLEMS INTRODUCING ERGONOMICS INTO THE DESIGN PROCESS

In 1988, Pheasant suggested five common fallacies about ergonomics, which enable designers to think they can 'successfully' design products without such information:

- This design is satisfactory for me — it will therefore be satisfactory for everybody else.
- This design is satisfactory for the average person — it will therefore be satisfactory for everybody else.
- The variability of human beings is so great that it cannot possibly be catered for in any design — but since people are so wonderfully adaptable it doesn't matter anyway.
- Ergonomics is an excellent idea. I always design with ergonomics in mind — but I do it intuitively and rely on my common sense so I don't need tables of data.
- Ergonomics is expensive and since products are actually purchased on appearance and styling, ergonomics considerations may be conveniently ignored.

The work of Hasdogen (1995) substantiates Pheasant's view in revealing that ergonomics human models are still not widely used in the product design industry. She found both designers and ergonomists critical of physical and psychological human models; their appropriate use being sometimes clumsy and awkward as a consequence of the poor design of the tool. Ergonomists, however, are still willing to use the data provided as a starting point for design (accepting its limitations) while designers tend (because of these limitations) not to use the data and to base their designs on their internalised knowledge/experience (as described in the fallacies above) which they assume to be correct. The input of ergonomics to the design process appears sometimes to pose a difficult problem, in particular the interface between designers and ergonomists. This is essentially a communication problem which manifests itself in three ways:

- Communication of ergonomics information at an inappropriate point in the design process.
- Communication difficulties between ergonomists and designers/engineering designers caused mainly by educational and practice differences.
- Communication of ergonomics information and data in an inappropriate fashion, by ergonomists.

2.4 COMMUNICATION AT AN INAPPROPRIATE TIME

Ergonomics data and information input into the design process is often not planned and may occur only at a point at which a 'human interaction' problem is identified. This may be very early, when there are prototypes for people to explore, or not until the product reaches the market place. In the automotive industry, for example, ergonomics tends to be an engineering function and as such happens after the aesthetic design is 'signed off' (see Tovey 1992). Prior to this sign off, a basic 2D package including some 'hard points' (established engineering or ergonomics 3D data points) which must be incorporated into the design, may be all that has been provided to the designer. Unfortunately, this often means that a late ergonomics input can become very expensive in both time and financial terms. For example, any consequent change in the design specification may require expensive and lengthy tooling changes, with all the accompanying legal requirements, material strength checks and so on. From a designer's viewpoint it may require unwelcome changes to the aesthetic of the product. Designers tend to become committed to their aesthetic at a very early stage in the design process and can feel very resentful about subsequent forced changes to their design.

The corollary of this problem is, of course, how do ergonomists provide input at an appropriate time. As concurrent engineering practices become increasingly established and new technologies, such as Computer Aided Design and Virtual Reality, are introduced into the aesthetic as well as engineering design process (Porter and Kehler 1996; Botley, Porter and Newman 1997) there are good opportunities for providing an ergonomics input into the design process at the early stages. However, it is likely that design teams also need to adopt new ways of thinking and procedures which encourage this opportunity.

A problem identified by the authors, through teaching experience, is germane to this issue. The discipline of ergonomics has produced many checklists against which it is possible to evaluate an existing product against ergonomics criteria. However, there is little evidence of the same investment of effort to develop tools that can be used to identify ergonomics issues during the concept phase of the design process. Woodcock and Galer Flyte (1997) have developed the Automotive Designers Ergonomics Clarification Toolset (ADECT), a design tool which provides a structure for the systematic investigation and recording of ergonomics issues in the early stages of the automotive design process. This approach was based on HUFIT, *Human Factors Tools for Designers of Information Technology Products* (Allison *et al.* 1992). These tools, however, go beyond the simple checklist approach to include methods and data and their use could be perceived as too onerous by the intended user population of designers. Burns *et al.* (1997) have identified that engineering designers perceive there to be a high cost involved in finding appropriate information among the mass of that potentially provided. They describe designers as "conservative searchers, probably as a consequence of adapting to domain constraints". Their studies were conducted with a subject group of engineering students, arguably better equipped through their education for this type of problem solving activity and this type of data than many arts based design students. The development of such design tools can only improve the current situation but the production of them and the data to which they refer needs careful consideration and user testing. The firm conviction of Burns and Vicente is that handbooks are not the answer to the problems of communication of ergonomics information to designers at any point in the design process.

2.5   COMMUNICATION DIFFICULTIES WITHIN THE DESIGN TEAM

Burns and Vicente (1994) investigated the use of ergonomics information by design engineers in the nuclear industry and the designers were found to be reluctant to use the information. As stated above, it was perceived to be of high cost to obtain and of low value. This was in part because of the difficulties of accessibility, and the type of available information being perceived as unsuitable for their needs and requirements. This may be a consequence of the different design approaches taken by engineers, designers and ergonomists. A number of these were recently identified by Porter, Case and Freer (1996). These differences probably occur as a consequence of innate ability, education and the real world practice of the different disciplines.

Dealing with variability

Engineering education and practice dictate an approach which is to design out variability by the selection of high quality materials and methods of manufacture. As a consequence, the statistical mean of an attribute of the product, be it a dimension, weight, strength etc., is a very good predictor of all products being produced at that time. The statistical mean is also the best predictor for people variables, although human variability is so great that the mean is of considerably less interest for design purposes. Put crudely, designing for clearances, reach and strength using the statistical mean for the appropriate variable (e.g. sitting height, arm reach, grip strength) will result in up to half of the intended population being 'designed out' for each variable.

A classic example of the importance of understanding human variability is in the design of a receptacle for medical products which needs to be openable by adults but should be difficult to open by small children. One of the most common solutions is to require large forces and a high degree of manipulative and cognitive abilities. Unfortunately, this also designs out a large number of elderly people who may be the prime users of such a product.

Jurisdiction and communication

Stylists and designers are particularly adept in communicating their ideas as they rely chiefly upon the visual image. As the major influences for a design are achieved at the concept stage (Tovey 1992) then such people also have a wide jurisdiction of influence. For example, their sketches of handles and seats may be stylish but may be designed with little attention to ergonomics principles (Hasdogen 1995); however, they can form the basis for the production specification.

Engineers are also able to communicate effectively with each other through engineering drawings and technical analyses. Ergonomists typically contribute to design by providing data concerning human characteristics and by providing evaluative data concerning issues such as discomfort, usability and safety. These inputs do not directly influence the design unless the solution can be sketched or dimensioned accurately. Consequently, the ergonomist has traditionally had to rely upon the support of the other team members in order to incorporate the ergonomics specification. Haslegrave and Holmes (1994) also identified this as a problem and suggested that we must look towards giving a better education to ergonomists, engineers and designers, in the understanding and use of each others' methods, in order that they may work more effectively together.

While it is unrealistic to expect that ergonomists should be able to sketch, for example, in the same way as designers do, the fact that they rarely have this skill can compromise their ability to successfully communicate their input, and for designers and engineers to easily use it. This problem is being addressed educationally with an increasing number of ergonomists being employed to teach on product design courses, giving designers and engineers a better appreciation of ergonomics and encouraging ergonomists to develop more effective means of communication. It would be equally sensible for designers to be involved in the education of ergonomists, to help ensure that their training involved communicating with designers.

2.6 COMMUNICATION IN AN INAPPROPRIATE FASHION

There is an abundance of ergonomics information available to the designer, the most commonly used being anthropometric data. The information is available in tabular, pictorial/diagrammatic form (e.g. Pheasant 1988) and also as design recommendations (e.g. Dreyfuss 1967; Diffrient, Tilley and Bardagjy 1978; Diffrient, Tilley and Harman 1981a and b; Panero and Zelnik 1979; Tilley 1995; Woodson 1981). They are all designed to provide information which allows designers to base their designs on apparent fact; the size of particular body dimensions of very small and very large females and males. It is interesting to note that many of the most successful forms of ergonomics data used by particular disciplines, have been produced by individuals from within that discipline who have been able to tailor the product to their requirements. For example, Pheasant for ergonomists, Tilley for designers, Panero and Zelnik for architects and Woodson for engineers.

There are both content and presentation differences in the various forms of data. The tabular and pictorial forms both provide data from which the designer is responsible for selecting appropriately for his or her requirements, i.e. in terms of age, gender, nationality, clothing etc. as well as choosing the correct body dimensions for the particular design. Design recommendations (e.g. for seat width) are another form of data where various assumptions have already been made, on behalf of the designer, by the writer of those recommendations. Consequently, the recommendations are often either too general or too specific. For example, a general recommendation to allow or enforce the maintaining of a straight wrist in hand tool design (good ergonomics practice) when the context in which the tool is to be used is not known or considered, may mean that the design of a tool which facilitates this may be imposing another postural constraint which is going to provide a greater long-term health risk. In some cases (where a great number of assumptions have been made by the writer of the recommendation) the guideline is much more specific than that which the designer requires to specify a design. In an ideal world the designer would be encouraged to think about and specify the assumptions he or she wanted to make, in an interactive way, before the recommendation was produced. The issues described can make the use of the information confusing and it is possible to see how 'inappropriate' use might occur.

Ergonomics information is now available in similar forms to those described above, as computer databases (e.g. *ERGOBASE* and *PEOPLE SIZE* — a relatively new package) and also as 3D human modelling ergonomics evaluation tools (e.g. SAMMIE, JACK, RAMSIS, Porter *et al.* 1995). The latter systems allow the possibility of modelling prototype designs and evaluating the designs with human models of varying body dimensions to examine issues of fit, reach, vision and posture, as already explained for the lightweight sports car. This can remove the necessity for extensive mock-up stages in the design process and such packages are ideal for exploratory investigations in the early stages of design.

A recent study conducted in Coventry School of Art & Design (Porter 1995) showed that industrial design postgraduates were reluctant to use readily available computerised data sources (ERGOBASE and PEOPLE SIZE). The student designers found the utilisation of the data difficult for a number of reasons:

- The forms in which the data are presented. It has been noted, by academic and library staff alike, that design students prefer to use the more visual forms of either traditional anthropometric data or recommendations (e.g. HUMANSCALE which is particularly popular because it provides visual recommendations). The students find the terminology of anthropometry, derived from an anatomical root, obscure and not all the tools clearly or simply address this issue. PEOPLESIZE is a major step forward in this respect. It is a picture driven system (with users choosing body parts for which they require dimensions) rather than having to know an anatomical term. It is much better liked by design students because of that. However there were still complaints about the aesthetic qualities of the illustrations, although the recently released version 2 has addressed this issue. Designers are visual/spatial people (Muller 1989; Tovey 1992) and would like the information they use to be visually appealing. This would also allow them to take the visual information and incorporate it directly into design research reports and package drawings. In an ideal world a package such as PEOPLESIZE would produce an output of sufficient aesthetic appeal that the designer would feel able to include it in the design work and consequent communication with others.

- It is often not possible to source the data for particular dimensions. Some of the information which is required in order to use anthropometric data successfully is not available or is difficult to access (e.g. age, nationality, specific nature of the measurement, the specific posture adopted while the data were being collected and clothing type) in all three packages. This makes them not as useful as some of the traditional sources on which they were based.
- Such findings can clearly interfere with the process of successful design for human use and also suggest that ergonomists must carefully consider the presentation aspects of the tools they believe designers should use.

2.7 CONCLUSIONS

Whilst it is easy to argue that the application of ergonomics information and data to the design of safe and successful products is essential, there appear to be a number of reasons why this may not always happen as a matter of course in the design cycle. The focal point of the reasons have their bases in communication. Ergonomics data and information are often communicated at a late and expensive point in the design cycle, educational and practice differences between designers and ergonomists make communication of the data and information sometimes difficult and it is often not conveyed in a form which is appropriate to the designer. In essence the *when* and *how* of ergonomics information and data are not always optimised. This will obviously hinder the process of successful design for human use.

It is possible for these problems to be overcome as the case study shows. The ergonomics input happened as the very first stage of the design process. SAMMIE, the ergonomics design tool used, proved to be an ideal one. It provided information for the designers in a visual form that they could relate to, it provided dimensioned data for both engineering and design purposes and was able to cross language barriers and computer systems. The case study also involved the input, as required, of an ergonomist and their expertise. However, as suggested earlier in the chapter this is often not an economic reality in the 'real world' and a major challenge for ergonomists is to provide tools for designers at all levels (not only sophisticated computer aided tools) which are accessible, understandable, and also encourage the use of ergonomics during the early stages of the design process.

2.8   REFERENCES

Allison, G., Catterall, B., Dowd, M., Galer, M.D., Maguire, M. and Taylor, B., 1992. Human factors tools for designers of information technology products. In *Methods and Tools in User Centred Design for Information Technology*, M.D. Galer, S.D.P. Harker and J. Zeigler (Eds), Elsevier Science Publishers, The Netherlands, pp. 13-42.

Botley, P.J., Porter, C.S. and Newman, R.M., 1997. *DMU — a Framework for Concurrent Design.* Paper accepted for 13th National Conference on Manufacturing Research, September 1997, Caledonian University.

Burns, C.M. and Vicente, K.J., 1994. Designer evaluations of human factors reference information. In *Proceedings of 12th Triennial Congress of the International Ergonomics Association,* **4**, Ergonomics and Design, Toronto.

Burns, C.M., Vicente, K.J., Christofferson, K. and Pawlak, W.S., 1997. Towards viable, useful and usable human factors guidance. In *Applied Ergonomics,* **28**, 5/6, pp. 311-322.

Department of Trade, Consumer Safety Unit, Millbank Tower, Millbank, London, *The Home Accident Safety System.* Statistics published yearly.

Diffrient, Tilley and Bardagjy, 1978. *Humanscale 1/2/3,* MIT Press, Cambridge, Mass.

Diffrient, Tilley and Harman, 1981a. *Humanscale 4/5/6,* MIT Press, Cambridge, Mass.

Diffrient, Tilley and Harman, 1981b. *Humanscale 7/8/9,* MIT Press, Cambridge, Mass.

Dreyfuss, H., 1967. *The Measure of Man: Human Factors in Design.* The Whitney Library of Design, The Architectural Press, London.

Hasdogen, G., 1995. The nature and limitations of user models in the household product design process. *Design Studies,* **17**, 1, pp. 19-33.

Haslegrave, C.M. and Holmes, K., 1994. Integrating ergonomics and engineering in the technical design process. In *Applied Ergonomics,* **25**, 4, pp. 211-220.

Henry Dreyfuss Associates, 1993. *The Measure of Man and Woman.* The Whitney Library of Design, N.Y.

Muller, W., 1989. Design discipline and the significance of visuo-spatial thinking. In *Design Studies,* **10**, pp. 12-23.

Panero, J. and Zelnik, M., 1979. *Human Dimension and Interior Space.* The Whitney Library of Design, The Architectural Press, London.

Pheasant, S., 1988. *Bodyspace.* Taylor & Francis, London.

Pheasant, S., 1995. A forseeable risk of injury. In *Contemporary Ergonomics,* S.A. Robertson (Ed.), Taylor & Francis, London, pp. 2-13.

Porter, J.M., Porter, C.S. and Lee, V.J.A., 1992. In *Contemporary Ergonomics,* E.J. Lovesey (Ed.), Taylor & Francis, London, pp. 262-267.

Porter, S., 1995. The use of ergonomics software in the education of industrial designers. In *Proceedings of the 2nd National Conference on Product Design Education.*

Porter, J.M., Case, K. and Freer, M.T., 1996. SAMMIE: a 3D human modelling computer aided ergonomics design system. *Co-Design,* 1996, pp. 62-73.

Porter, J.M., Case, K., Freer, M.T. and Bonney, M. C., 1995. Computer aided ergonomics and workspace design. In *Evaluation of Human Work,* J.R. Wilson and E.N. Corlett (Eds), Taylor & Francis, London.

Porter, J.M. and Gyi, D., 1998a. Exploring the optimum postures for driver comfort. *International Journal of Vehicle Design,* **19**, 3, pp. 255-266.

Porter, J.M. and Gyi, D., 1998b. Seat pressure measurement technologies: considerations for their evaluation. *Applied Ergonomics,* **27**, 2, pp. 85-91.

Porter, C.S. and Kehler, T., 1996. TR102, *Restricted Circulation Document,* EU funded INSTANCE Consortium document.

Porter, S. and Saunders, S., 1996. The use of computer aided styling tools in the automotive design process. *Co-Design,* **8**, 1996, pp. 56-61.

Tilley, A., 1995. *The Measure of Man and Woman, Henry Dreyfuss Associates.* The Whitney Library of Design, The Architectural Press, London.

Tovey, M.J., 1992. Intuitive and objective processes in automotive design. *Design Studies*, **13**, 1, pp. 23-41.

Woodson, W.E., 1981. *Human Factors Design Handbook.* McGraw-Hill Book Company, London.

Woodcock, A. and Galer Flyte, M.D., 1997. The development of computer based tools to support the use of ergonomics in automotive design. In *Proceedings of IEA 1997*, **2**, Tampere, Finland, pp. 124-127.

CHAPTER THREE

Product Evaluation Methods and Their Importance in Designing Interactive Artifacts

V. POPOVIC

School of Architecture, Interior and Industrial Design, Queensland

University of Technology, GPO Box 2434, Brisbane 4001, Australia

ABSTRACT

This chapter discusses the methods and techniques to be used for an ergonomics evaluation of products, product interfaces and systems. In this context evaluation is seen as an interactive part of all stages of the design process. It plays an integrative role in design and is concerned with ensuring a high degree of likelihood of user's acceptance. Traditional and new evaluation methods and tools, such as task analysis, checklists, TA (talk/think aloud) protocols and CAD simulation, are addressed. The paper also highlights the issue of user–artifact–designer interaction in order to reinforce the need for a consistent user-centred design approach, from concept generation to artifact/system development and production.

3.1 INTRODUCTION

To succeed, a product or system must provide satisfactory interaction with its user/customer on both a functional and a cultural level. Manufacturing companies are competing on national and international levels to achieve a competitive edge in the market. This creates demand for faster product development and production. Product quality refers to the performance, overall design and interface design of the product or system, the manufacturing process, and the product life cycle. This means that better design, developed in detail and based on applied research during the design and development process, plays a significant role in the competitiveness of a company.

3.2 USER-ARTIFACT-DESIGNER INTERACTION

Design is a prediction concerned with how things ought to be. It is aimed at changing an existing situation into a preferred one. Designers attempt to predict the behaviour of a product and its users through their knowledge and expertise. To solve a problem the human has to understand it. To achieve this, designers have to understand what the knowledge structure is that humans have regarding products and systems and their contextual environment; the environment in which products are used. However, designers still operate in their traditional role (that is, in a professional client relationship). The designers receive the client's brief in which needs and wants are specified, and design a product outside its contextual environment by predicting the behaviour of a product and its users on the basis of their knowledge as experts, or from personal experience. The outcomes of this are products/systems that do not respond to user's expectations. They are designed 'for users' but not 'with users'. Designers, via market research information, interpret the user's needs or employ themselves as user stereotypes. This causes problems at the interface of the interaction between users and products (Norman 1986, 1988, 1993). However, the traditional role of designer-client interaction is changing into a more complex one, which means that in-depth research is needed in order to design better and more valuable products/systems that will respond to contemporary demands. It will require designers to apply more sophisticated knowledge in order to respond to market demands and generate users' satisfaction. Therefore, designers need to understand how they design, what and for whom they design, and how to implement innovation into their designs. To the greatest extent possible, they have to integrate design knowledge and domain-specific knowledge about the product users. They must understand the product users and their knowledge and experience of the products they operate (or may operate).

The ideal requirements of design are to achieve the highest level of compatibility between user's and designer's models. To achieve this, designers should begin designing with good knowledge of the users, and include users as a part of the project team. Before establishing any framework for user representation, it is necessary to outline briefly the state-of-the-art features of user designer interactions in order to qualify the gap between them. Users and their needs are still interpreted by a designer, an interpretation that is based on an incomplete user stereotype. Alternatively, designers see themselves as 'users', a view that causes a range of conflicts when the users (both novice and expert) eventually come to interact with products. Hammond, Jørgensen, MacLean, Barnard and Long (1984) interviewed designers about their 'theories' of users. In general, designers found it difficult to express a theory about users. Some based their judgements on intuition and employed themselves as a user model by applying their own experiential knowledge (Visser 1996). One of the major criteria during the design process is that the artifacts maintain consistency and performance — from initial concept to distribution in the marketplace — from the end users' point of view. Users' needs and characteristics ought to be considered during all stages in the design and development of an artifact: all artifact design should be user-centred. It is well known that the design stages of a product cannot be separated from the overall cycle of its development and production. It is essential to remember this when we talk about manufacturing products by whatever process of mass-manufacture or advanced manufacturing systems, since methods of production have strong implications for product design.

With the development of technologies, designers have unlimited possibilities for designing a product, its interface and functions. The more varied the functions, the greater the need for a product interface to be easy to learn and operate. However, it seems that designers, instead of applying a user-centred design approach, are still aiming toward a designer-centricity that is alien to users (Hammond, Gardiner, Christie and Marshall 1987). The whole contextual environment of artifacts affects levels of interaction and artifact use. The ways they are used and level of interaction that is to be achieved depend on users' tasks and expertise. Artifact design can affect a user's task and performance. If an artifact is designed without task analysis, or is not user-centred, then it will impose difficulties on users (Norman 1988). This distant view of an artifact and its contextual environment generates conflicts between users and artifacts.

End-user satisfaction is more and more becoming a standard requirement in all artifact design, from technologically interactive devices to complex interfaces and systems. Kato (1986) referred to Moran's paper in which he called for the establishment of an "applied psychology of the user" as a basis for understanding human cognitive activities. In the cited reference it was pointed out that it is unacceptable for designers to view themselves as users, because the two groups differ greatly in their approaches. This is one of the reasons that designers' concepts and users' expectations and understanding of systems continue to differ. Designers and users form different mental representations of the tasks and the interface, which are therefore often mismatched. This suggests that users' knowledge is different from designers' knowledge (Jørgensen 1990). Designers' visions and people's needs differ greatly (Bannon 1986). In his paper, Jørgensen (1990) stated that the role of cognitive psychology is to study human cognitive processes, while in the area of user interface design, the purpose is to identify errors. He emphasised the view that 'how the users' cognitive processes take place is of no interest to designers (but has of course implicit implications for them)". If designers want to offer the users good and helpful systems then 'cognitive problems' that prevent them from successfully interacting with artifacts or systems should be recognised "using strategies that are actually used by designers to solve them" (Visser 1992).

The point has already been made that designer–client relations should change from a traditional to a more complex one in which artifact users are seen as clients. It is apparent that designers are expected primarily to satisfy their clients. Clients are interested in costs and profits, but they are not the ones who are going to use the products. Once the products are sold, the manufacturers have no further direct interest beyond reliability and repair costs. However, if users reject products that are on the market, then product sales will be affected. The interest in users is becoming more acute, and increasingly, product usability is being explored by designers. This is done with limitations, as product designers are more concerned to make a product 'look different', and this may bring an additional problem to the human users. This leads to the debate about designers' responsibilities.

Designers have their codes of conduct which are adopted by their professional associations and until very recently, users were not mentioned in the professional practice codes. However, in 1995 the International Council of Societies of Industrial Design (ICSID) revised its codes, and Article II now states that "the Industrial Designer will advocate and thoughtfully consider the needs of all potential users including those with different abilities such as the elderly and the disabled". This supports the premise that designers are required better to understand the users of artifacts.

The human user is an "information processor" whose behaviour is not easy to predict (Sutherland 1994). This is another reason why designers need to understand their users and be able to model users' tasks during the early stages of the design process. It requires the designer to predict what users will know or be able to learn. Designers are supposed to map psychological principles into their design decisions.

3.3 EVALUATION METHODS AND DESIGN PROCESS

The most innovative phase of design process is its conceptual phase, in which most decisions are made. With advanced product development and manufacturing more detailed product concepts are needed. This means that in this stage of the design process designers need to predict users' behaviour and the operation of products or systems. One of the major directions during the design process is *that the products should manifest the end users point of view, from initial concept to their distribution to the market place*. This means that user constraints should be included into the design project from its initial stage and followed throughout the project consistently (Popovic 1997; Popovic 1983). In order to achieve this a designer must have a body of knowledge about users and their behaviour which can be obtained from (a) research, (b) evaluation of same products/systems, (c) evaluation of related products/systems, and (d) evaluation of predicted products/systems. Therefore, evaluation is seen to be the part of the design process that interacts with all its stages. It occurs during the whole design process, which is contrary to some design methodologies that consider evaluation at the end of the particular project stage. Because of potential weaknesses in design concepts and their consequences, evaluation should be reinforced in the early stages. The nature of a design project determines which kind of tools, methods, strategies and knowledge is required (Tables 3.1 and 3.2).

Table 3.1 Common evaluation tools.

TOOLS	PURPOSE	DESIGN PROCESS STAGES
CAD simulation and Virtual reality (VR)	To evaluate design and its perceived use.	Concept development stages.
Mock-up evaluation	To evaluate product usage with users' participation.	Concept development stage.
Prototype evaluation	To verify a design outcome under real conditions.	Different stages of the design process.

Table 3.2 Common evaluation methods and techniques.

EVALUATION METHODS/TECHNIQUES	PURPOSE	DESIGN PROCESS STAGES
Checklists	To define operations of a product/ system and identify users' needs.	Early stage of the design process and field test.
Focus group	To identify user issues and their importance.	Any stage of the design process.
Interviewing users	To identify users' needs.	Any stage of the design process.
Observation techniques	To define dynamics of the artifact/ system/environment.	Final design stage and field test.
Protocol analysis	To evaluate a design, users' expertise levels and understand users' concept of products.	Any stage of the design process.
Task-analysis	To define and evaluate operational procedures of human/product/ system.	Concept development stage, final design stage and field test.

They are applied for assessing product/artifact usability as separate techniques or in combination (Jordan 1998; Jordan *et al.* 1996). The selection of the appropriate method will depend on design goals — which design constraints have to be evaluated. For example, to identify users' needs a designer may decide to select interviews or checklist evaluation or to conduct focus groups; to understand users' tasks and their underlying knowledge, task and protocol analysis can be used; to assist in the evaluation of usability at a design concept stage, CAD simulation techniques can help.

3.4   EVALUATION TOOLS, METHODS AND TECHNIQUES

The most common evaluation tools, methods and techniques are listed in Tables 3.1 and 3.2. They are seen to be the methods that will help designers to gain better knowledge about the artifact users and at the same time evaluate useability of their designs. All these techniques can be applied through the different stages of the design process. Their brief descriptions are as follows:

EVALUATION TOOLS

CAD simulation models and Virtual Reality (VR)

CAD simulation models can be used to evaluate design and its perceived use during the *different stages of design process*. These are representations of artifacts or artifact interfaces that can be simulated, assessed and changed during the design process. This is a method easily applied and it is useful to evaluate the ideas and initial concepts. It provides an opportunity to test a number of design alternatives before the final decision is made. However, its disadvantage is that a lot of decisions remain uncertain, as they have to be verified through an evaluation of a prototype. CAD simulation is meaningful when a user's perception of the artifact is required or when the design team needs to study an overall design. In this case any valid usability judgement is difficult to make (Jordan 1998).

The development of Virtual Reality (VR) and its wider availability are opening new avenues for its application in the evaluation of a product design and its usability. By placing a virtual or human user into VR it may be possible to evaluate artifact usability in its context in all design phases. This may have many advantages, such as evaluation of a truck driver workplace design, or design in an extreme environment such as space projects, where human-product interaction can be assessed before construction. The design tools that are emerging are (a) integration of CAD and VR, (b) immersive technologies and (c) telepresence (Kawrowski *et al.* 1997).

Mock-up evaluation

Mock-up evaluation helps in assessing product usage during concept development stage. It is the method that involves users' participation and usability testing. It can incorporate different methods and techniques. It also helps to analyse and evaluate how an artifact may fit into its contextual environment of use. There are different mock-up simulation techniques applied during the design process such as (a) simulation of design features, (b) simulation of interface and proposed interaction and (c) simulation of an artifact's contextual environment. A range of useability evaluation methods can be applied — from protocol and task analysis to interviews and checklists. For a detailed examination of the possibilities and weaknesses of model evaluation see Rooden and also Kanis in this volume.

Prototype evaluation

Prototype evaluation helps to verify a design outcome under real conditions. It is the most effective method of assessing the usability of an artifact or system. It encompasses multiple usability methods and techniques and its main advantage is that it can be done in field. The aim is to get information and understanding of how an artifact/system operates in its context; the environment in which the artifact is going to be used. Its disadvantages are complexity and the cost involved during evaluation as it is at the end of the design process. With the emergence of rapid phototyping it is possible to evaluate artifacts earlier during the design and development process. This helps to identify design deficiencies and improve product usability.

EVALUATION METHODS AND TECHNIQUES

Checklists

Checklists are used to help to define operations of a product/system and identify users' needs. They also help to identify the deficiency of similar products. If designed properly they can provide reliable information about the user profile. They are good methods which may provide an overview of information about artifact usage. Checklists do not provide rich data about the users' experiences (Jordan 1998) of an artifact and their interactivity. They are also used to develop recommendations or 'score human factors variables' (Vianen *et al.* 1996). Checklist may be long or short and can be taken in conjunction with other evaluation methods such as observation or task analysis.

Focus users' group

A focus users' group helps to test the projected use of the product or system. The most common approach is that users are asked to discuss issues related to the usability of the proposed designs. They can be grouped according to their level of expertise such as: naive, novice, intermediate or expert users. During focus sessions, different techniques may be used. This method helps by identifying the issues that are important for the users but were not taken into consideration by designers. The focus groups consist of facilitators (leaders) and a number of participants. They are not rigidly structured, in order to allow participants to direct the discussion. The role of facilitator is to ensure that all participants have an opportunity to voice their opinions. It is appropriate for the facilitator to have some questions ready in order to make prompts if necessary. This is to control participation in case the discussion goes outside the area of interest. Focus groups have their disadvantages as they rely on the interpretation of data. There are possibilities for inconsistency if some members of the group are dominant and the outcome reflects the opinion of an individual rather than the group. Therefore, the role of the facilitator is very important in managing the group dynamics. The advantage of the method is that it can be applied at any stage of the design process. It supports the discussion about the artefact's design and usability (Jordan 1998). It also helps the designers to evaluate the artifact they design before design is finalised. See Trathen *et al.* (Chapter 15) for a description of a case study using focus groups.

Interviewing users

Interviewing users helps to identify users' needs and better understand their culture and the contextual environment in which artifacts are going to be used. It gives designers better knowledge about users' acceptability of design concept. It also clarifies users needs. The interviews can be unstructured, semi-structured and structured (Jordan 1998). In unstructured interviews the investigators may ask open-ended questions that will clarify the ideas and issues that are important to the users. Semi-structured interviews provide a more systematic approach and the structured interviews ask subjects pre-set questions. The advantages of interview methods are that they are versatile and can be applied at any stage of the design process. However, their disadvantage is in the administering cost, as the investigator is required to be present. The presence of an investigator may influence the answers, as the respondent may wish to give what they think may be the 'right' opinion. See also Kanis (Chapter 4).

Observation techniques

Observation techniques are applied to help in defining the dynamics of the product/system/ environment. They can provide insight into difficulties users have while interacting with artifacts or systems. This is done directly or via analysis of video and audio recording of the artifact in use. This method can be done in conjunction with task and protocol analyses. The information received is qualitative but it can be used to quantify users' interaction further. The major problem is to know how to observe and what to observe. Baber and Stanton (1996) suggested guidelines to be used prior to the observation taking place. These include definitions of the following: (a) aim of the observation study, (b) scenario in which the product will be used, (c) type of data to be collected, (d) presentation of data, (e) time available for observation, (f) recording tools and (g) constant evaluation of observational stages.

Protocol analysis

The protocol method or the think-aloud (TA) method is applied to studying human behaviour in different domains of expertise. Ericsson and Simon (1993) first described this method. It was expanded by van Someren *et al.* (1994). The amount of research utilising protocol analysis has been growing in recent years especially when computer simulation became available. The protocol method is widely accepted in the research community. Its data are unstructured and very rich, and flexible analytical methods can be used. In general, verbal and video recording of a user is made. Transcripts are made, segmented, interpreted and analysed. It is required that verbal protocol data should be put in an appropriate framework in order to get the best understanding of the analysed activity. There are some criticisms about giving verbalisation concurrently with the cognitive processes (Bainbridge 1990; Rooden, Chapter 14). Distortion may occur if a person does the work in a non-verbal way and is not aware of the verbalisation task as part of the work is done automatically. It is possible that verbal reports become distorted as task performance and verbal representation become incompatible. However, Berry and Broadbent (1984) investigated the relationship between cognitive task performance and reportable knowledge associated with it. Their experiments examined effects of task experience, concurrent verbalisation and verbal instructions. They found that concurrent verbalisation did not have any effect on task performance. Despite the criticism, think-aloud protocol is found to be very useful for analysis of interface design and human computer interaction. The method can help the designers to get a better understanding of the principles behind their concepts. It is applicable at any stage of the design process.

Task analysis

Task analysis is used to evaluate products and users' interactions with them via their interface and to assess their usability. In this context task analysis refers to overall users' activity. Its most common forms of representation are diagrams or charts. The methods and techniques are different for specific applications such as workplace design, medical equipment design, interface design, or knowledge elicitation. Task analysis of different products, conducted by the author but not reported here, identified the discrepancy between users' and designers' concepts of products or systems. The outcome of this analysis is used as constraints for new product designs and their concept evaluation.

Task analysis and protocol methods can complement each other. They can be used concurrently. When distortion of a TA protocol may occur, task analysis can clarify the sequences of operation. Video recording of the performance during concurrent verbalisation can be analysed to determine task components. In this situation task analysis techniques are used to support protocol data.

3.5   CONCLUSION

Designers and users are the agents of change: designers in their own domain as they design and bring to life artifacts using their knowledge and expertise and users in their own domain of expertise; the use of the artifact. Every interaction with a product interface generates changes by the user whose own goals and intentions change. By changing goals and intentions, the action and its outcome both change. The user is an agent who directs the complete interaction (Laurel 1986). This will become more apparent as technologies develop and interactive interfaces become easier to use. However, there are constraints built into a user's interaction and they are usually part of design constraints of the system. They keep users' activities within boundaries, but at the same time challenge users to enjoy different levels of interaction.

It was stated earlier that end-user satisfaction is increasingly becoming a standard requirement for all of the products and systems we design and use. Kato (1986) referred to Moran's paper in which he pointed out that is unsuitable for designers to use themselves as models of the user population. They are different. This approach is still common in many design domains where designers and design educators introduce an approach to design as being 'an understanding of yourself' or of their own ideas based on personal experience. This is one of the reasons that designers' concepts and users' expectations and understanding of products and systems differ. This suggests that users' knowledge is different from designers' knowledge. (Jørgensen 1990). Therefore, it is very important that designers understand that they must take into consideration many different factors, and study users' needs, expectations, concepts, behavioural patterns, culture, and the environment in which the products will be used, in order to ensure users' acceptance and enjoyment.

3.6 REFERENCES

Baber, C. and Stanton, N., 1996. Observation as a technique for useability evaluation. In *Useability Evaluation in Industry*, P. Jordan, B. Thomas, B. Weerdemeester and I. McClelland (Eds), Taylor & Francis, London, pp. 85-94.

Bainbridge, L., 1990. Verbal protocol analysis. In *Evaluation of Human Work*, J.R. Wilson and E. Nigel Corlett (Eds), Taylor & Francis, N.Y., pp. 161-180.

Bannon, L., 1986. Issues in design: some notes. In *User Centred System Design*, D. Norman and S. Draper (Eds), Lawrence Erlbaum, Hillside, NJ, pp. 25-29.

Berry, D.C., and Broadbent, D.E., 1984. On the relationship between task performance and associated verbalizable knowledge. In *The Quarterly Journal of Experimental Psychology,* **36A**, pp. 209-231.

Ericsson, K.A. and Simon, H.A., 1993. *Protocol Analysis Verbal Reports as Data.* The MIT Press, Cambridge, Mass.

Hammond, N., Jørgensen, A., MacLean, A., Barnard, P. and Long, J., 1984. Design practice and interface useability evidence from interviews with designers. In *Human Factors in Computing Systems*, A. Janda (Ed.), North-Holland, Amsterdam, pp. 40-44.

Hammond, N., Gardiner, M.M., Christie, B. and Marshall, C., 1987. The role of cognitive psychology in user-interface design. In *Applying Cognitive Psychology to User Interface Design*, M. Gardiner and B. Christie (Eds), Wiley, Chichester, pp. 13-52.

Kato, T., 1986. What "question-asking protocols" can say about the user interface. In *Int. Journal of Man-Machines Studies,* **25**, pp. 659-673.

Kawrowski, W., Chase, B., Gaddie, P., Lee, W. and Jang, R., 1997. Virtual reality in human factors research and human factors of virtual reality. In *From Experience to Innovation, IEA'97*, P. Seppala *et al.* (Eds), **2**, Finnish Institute of Occupational Health, Helsinki, pp. 53-55.

ICSID, 1995. ICSID *Codes of Professional Practice.* International Council of Societies of Industrial Design, Helsinki, p 2.

Jørgensen, A.H., 1990. Thinking-aloud in user interface design: a method of promoting cognitive ergonomics. *Ergonomics, 33*, 4, pp. 501-507.

Jordan, P., Thomas, B., Weerdemeester, B. and McClelland, I., 1996. *Useability Evaluation in Industry,* Taylor & Francis, London.

Jordan, P., 1998. *An Introduction to Useability.* Taylor & Francis, London.

Laurel, B., 1986. Interface as mimesis. In *User Centred System Design*, D. Norman and S. Draper (Eds), Lawrence Erlbaum, Hillside, NJ, pp. 67-85.

Norman, D., 1986. Cognitive engineering. In *User Centred System Design*, D. Norman and S. Draper (Eds), Lawrence Erlbaum, Hillside, NJ., pp. 32-61, 33.

Norman, D., 1988. *The Psychology of Everyday Things*, Basic Books, N.Y.

Norman, D., 1993. *Things That Make Us Smart*, Addison-Wesley, N.Y.

Popovic, V., 1983. A systematic ergonomics evaluation procedures and its application to ergonomics research. In *Proceedings of the 20th Annual conference of ESANZ "Ergonomics in the Community"*, T. Shinnick and G. Hill (Eds), ESANZ, Adelaide, pp. 15-23.

Popovic, V., 1997. Product evaluation methods and their applications. In *From Experience to Innovation, IEA'97*, P. Seppala *et al.* (Eds), **2**, Finnish Institute of Occupational Health, Helsinki, pp. 165-168.

Sutherland, S., 1994. *Irrationality,* Penguin Books, London.

Someran, M.W. van, Barnard, Y.F. and Sandberh, J.A.C., 1994. *The Think-Aloud Method: A Practical Guide to Modeling Cognitive Processes.* Academic Press, London.

Vianen, E. van, Thomas, B. and van Nieuwkasteele, M., 1996. A combined effort in the standardisation of user interface testing. In *Useability Evaluation in Industry*, P. Jordan,

B. Thomas, B. Weerdemeester and I. McClelland (Eds), Taylor & Francis, London, pp. 7-17.

Visser, W., 1992. Designers' activities examined at three levels: organisation, strategies and problem solving processes. *Knowledge-Based Systems*, **5**, 1, pp. 92-104, 103.

Visser, W., 1996. Use of episodic knowledge and information in design problem solving. In *Analysing Design Activity*, N. Cross, H. Christiaans and K. Dorst (Eds), John Wiley and Sons, Chichester, pp. 271-289.

CHAPTER FOUR

Design Centred Research into User Activities

H. KANIS

Faculty of Industrial Design Engineering, Delft University of Technology,

Jaffalaan 9, 2628 BX Delft, The Netherlands

ABSTRACT

In order to generate supportive insights for the design of domestic products, usage oriented research focuses on what people do and how and why they do it, in terms of perception, cognition, use-actions and effort. The observation of these user activities is cumbersome and difficult to achieve, as distinct from the identification of human characteristics which tend to combine good measurability with limited design relevance, acting primarily as boundary conditions. The obtrusiveness of observational research is discussed, together with the possibility of generalising observations, particularly in the case of simulations. Finally, it is argued that usage centred research in a design context thrives on a qualitative approach on the basis of small sampling, the strengths and weaknesses of which are discussed.

4.1 USAGE DEPENDENT FUNCTIONING OF EVERYDAY PRODUCTS

Figure 4.1 represents the functioning of a product operated by a user in order to achieve some goal, see Kanis 1998. In this representation, product functioning is seen as resulting from the co-occurrence of a product with use-actions in an environment. For a particular product operated under specified circumstances, this technical/physical process is set, given the actual use-actions. In their consequences, these use-actions can be conceived as reproducible by a robot. In their production, however, they emerge as human behaviour. Hence, user involved product functioning is exemplified by the key-role of user activities, i.e. perception, cognition and use-actions, including any effort involved.

In Figure 4.1, perception and cognition precede use-actions. Aside from the difficulty of disentangling perceptive and cognitive activities conceptually, the anticipation of use-actions by perception/cognition may be conceived in a variety of ways. Table 4.1 lists a number of combinations as established in different user trials. It shows that the observation of 'proper' use-actions, i.e. as envisaged in the design, does not in any way guarantee that these actions have been preceded by corresponding perceptive and cognitive activities. Similarly, difficulties in the actual operation of products may go along with perceptive and cognitive activities which are exactly as anticipated during the design process in order to activate proper use-actions. From a design point of view, it may make all the difference to know whether observed use-actions, particularly those indicating a deficient user product interaction, are preceded by, for instance, not noticing featural/functional cues, not understanding or misunderstanding such cues. Aside from these observations, use-actions can be seen to occur on different levels of perceptive/cognitive activity: compare the well-known distinction between skill-, rule- and knowledge-based product operation introduced by Rasmussen (1986). Hence the 'layered' presentation of perception/cognition in Figure 4.1, referring to various levels of involvement. In everyday product usage, the cognitive involvement tends to be low as smooth operation becomes the default in a continuous update of experience.

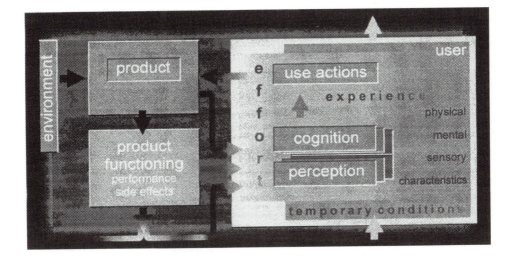

Figure 4.1   The functioning of a product operated by a user in order to achieve some goal.

Black arrows:	Product functioning as the result of a technical/physical process, i.e. as the outcome of the co-occurrence of a product with use-actions in an environment.
Grey-arrows:	Product functioning as negotiated by user activities, i.e. use-actions, cognitions and perceptions (the latter ones referred to by different layers, to indicate various levels of involvement, i.e. skill-based, rule-based and knowledge-based, cf. Rasmussen); experience: a psycho-motoric condition both activated and constituted in product usage.
Light grey arrows:	Additional paths via the environment for the possible effects of product functioning on users and vice versa; temporary conditions, including moods, experiences, e.g. feelings of inconvenience, of frustration, of enjoyment.

Table 4.1 User activities established in various user trials involving everyday products

Featural/functional cues* reported by subjects		Use-actions
Noticed?	Understood?	
No		• none (subjects 'stuck') • haphazard (right or wrong) • bypass, often at the cost of extra effort • operation as designed (cue(s) not needed)
Yes	No	• none (subject 'stuck') • haphazard (right or wrong) • bypass, often at the cost of extra effort. • operation as designed (cue(s) not needed)
Yes	Yes, but not as intended in the design	• 'erroneous'[1] (eventually 'stuck') • suboptimal, designed functionality not/only partly • activated
Yes	Yes, as intended in the design	• operation as designed • not as designed, alternative seen as more efficient • 'wrong'[1], 'mistake'[1] made (e.g. striking wrong key) • 'erroneous'[1] or suboptimal, as proper action is delayed

* Cues as collective term for product characteristics and feedback in product functioning in so far as these are directive of actual usage.

(1) Terms like 'error' or 'mistake' are assessments from an external (observational) point of view, for instance equivalent with ineffective, useless, which does not necessarily reflect the experience of the user involved.

4.2 THE PREDICTABILITY OF USER ACTIVITIES

Empirical research shows that user activities seem to be only loosely associated with human characteristics such as the sensory capacities, mental abilities and physical properties of human beings (Kanis 1998). The significance of these characteristics (see Figure 4.1 at the right) for the functioning of products in practice tends to be limited to setting boundary conditions for user activities (cf. Green *et al.* 1997), rather than serving as a base to make the activities and experiences of users predictable.

In addition, which 'human factors' effect the functioning of products in practice, largely escapes specification in current theoretical perspectives such as the information processing paradigm, ecological approaches or activity theory. To a considerable extent this is due to the level of abstraction in theoretical frameworks concerned (cf. Gelderblom 1998). In fact, activities of users are becoming less predictable the finer the level of description. Empirical observation seems an obvious way to keep track of these activities.

4.3   OBSERVABILITY OF USER ACTIVITIES, INCLUDING EFFORT INVOLVED

Use-actions

This type of user activity seems the most observable one, e.g. by video recordings. The omnipresence of use-actions in daily life contrasts sharply with the frequency of studies into these actions published in the area of Ergonomics/Human Factors (E/HF). This may be due to the lack of a clear insight into the impact of 'human factors' of users on the functioning of a product in practice (see Figure 4.1). What may also play a role is the explorative character and laboriousness of this type of study. They yield observations which are cumbersome to shape into categorical evidence, as distinct from experimental traditions in those disciplines of the social sciences, such as psychology, which are the main constituent of E/HF as a research area.

Perception/cognition

The identification of people's perceptive and cognitive activities, such as cues being noticed or understood, is not without difficulties. The two current ways to find out about what goes on 'inside' are retrospective revelation of perception/cognition and concurrent elicitation by 'thinking aloud'. Both approaches rely heavily on the assumption that the perceptual and cognitive activities involved can be shared on the basis of linguistic representations. For everyday products in particular, where usage tends to be automated on a skill-based level, the linguistic amenability of perceptive and cognitive activities may be limited, in as far as these activities are at issue at all. See Rooden (1998) for further comments, including the possible interference of 'thinking aloud' with carrying out a task and the unfortunate habit of subjects to fall silent for longer periods especially in case of difficulties, as well as the emergence of social desirability in explanations produced retrospectively. Nonetheless, a linguistic vehicle seems unavoidable in tracing perception and cognition, despite its potential ambiguity and the possible unwelcome side-effects of its application. As Rooden indicates, it is the combination of recording perceptive and cognitive activities with the observation of use-actions which makes sense. Since use- actions, as observed, cannot be denied, these actions may serve to underpin the plausibility of reported perception/cognition on the basis of the absence of conflicting evidence, that is, in so far as the data at issue add up to a coherent insight. In addition, the credibility of self-reports can be seen to be strengthened when these reports may be interpreted as revealing shortcomings in perceptive or cognitive abilities, i.e. running counter to a presumed proclivity of subjects to social desirability. For example, the somewhat embarrassing but consistent record of a subject in a user trial who did not use the automatic adjustment of the suction power to the type of floor covering when vacuuming, as she understood the term 'Auto' to be the appropriate position of the suction power regulation for vacuuming a car (Loopik *et al.* 1994).

Effort

Both self-reports and external observation may serve to record any effort, either in noticing, in figuring things out, or as physical exercise. Self-reports typically rely on the internal references of people, reflecting personal experiences. Obviously, such references are highly individual. An example is the assessment of control operations as 'effortless' by people suffering from arthritis, who were only able to manage by using two hands, while experiencing pain (Kanis and Van Hees 1995). This individual character of internal references also involves interpretative differences of terms, meant to trigger the 'same' references across people: think of words like 'effort', 'difficulties', 'problems'.

In addition to measurement of effort in physical terms, such as by an ergometer, external observation may be based on the time it takes subjects to carry out a particular task, the number of attempts, and the number of mistakes. In fact, such measures are substitutes, which, if averaged, may yield useful quantifications in so-called 'summative' evaluations, but which may have no bearing on any effort experienced personally. This is not to say that, ultimately, only self-reports would constitute the genuine observation of effort. For instance, in the example of the misinterpretation of the term 'Auto' (see above), the subject involved will not experience inconveniences or unnecessary effort, at least not as long as she is unaware of a more effective usage of the designed functionality.

Observability as a circular problem in design

The extent of the unpredictability of user activities confronts designers with a fundamental. Any new design should meet functional requirements in terms of its performance and (unwanted) side-effects (see Figure 4.1, lower left), and the effort involved (see the user-block in Figure 4.1 at the left). However, whether or not requirements will be violated depends to a considerable extent on factors which may only be observed once the new design is available. The only way to deal with this circular problem seems to include simulation of future user product interaction in an empirical approach to identify user activities. An obvious element in this simulation is whatever model or prototype is available in the design process. Note that the term 'simulation' is used in a narrow sense: in a design process, with the design progressed to some model, there is no reality to simulate, as the 'simulation' helps to create a future reality.

4.4   RESEARCH INTO (SIMULATED) USER ACTIVITIES

Figure 4.2 expands Figure 4.1 in specifying parameters at issu in the empirical observation of (simulated) user activities. In order to highlight the research context in Figure 4.2, the term 'subject' is used instead of 'user', and the term 'conditions' instead of 'environment' reflecting different grades of simulation, ranging from a field setting to laboratory conditions or virtual reality. For the 'product' in Figure 4.2, the references have been labelled in order to indicate its previous stages during the design process, i.e. as some model (e.g. conceptual, physical) or a working or partly working prototype.

Compared to Figure 4.1, the reference to instructions is an extra element in Figure 4.2. As distinct from users with immanent goals to be achieved by naturally operating a product, a subject somehow has to be tasked. Instructions may come in various ways, and, for instance, tend to be more explicit as to the steps to be taken sequentially, the less complete a design model. In addition, the character of an instruction may vary, in creating a use context or imposing specific tasks, see Vermeeren (1997). This diversity is reflected by the 'layered' references in Figure 4.1.

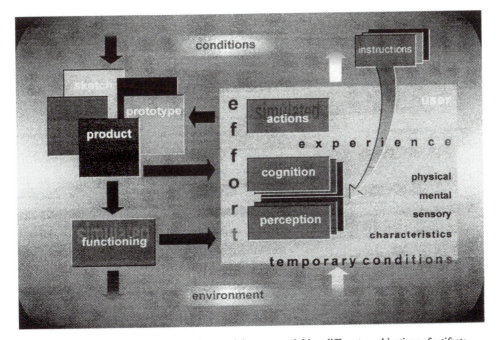

Figure 4.2 Parameters in design centred research into user activities: different combinations of artifacts (ranging from primitive models to fully functioning products), of reserach conditions including different types of instructions, and of possible future users varying in many different characteristics.

Obtrusiveness of human involved research

The more explicit the investigative context of human involved research (see Figure 4.2), the more obtrusive this research may be. And the more the obtrusiveness, generally, the more observations are contingent on a particular research context i.e. the more questionable generalisations may be compared to the ideal as reflected in Figure 4.1. This ideal would comprise the complete observation of user activities, including experienced effort (mental, physical) while the humans involved remain unconscious of being observed. As discussed above, this ideal is probably unattainable, since self-reports of the humans involved constitute the prime source of information, i.e. perceptive/cognitive activities and experienced effort. The ideal is even more remote in a design process using primitive models, which subjects can only 'operate' by verbalising their intended actions. Use-actions, in the actual operation of a functioning product, lend themselves particularly to unobtrusive observation. Since 'unobtrusive self-reports' is a contradiction in terms, unobtrusive observation of use-actions cannot coincide with simultaneously aired self-reports such as in thinking aloud.

Unobtrusiveness by research contexts and instructions

Self-reports, meant to be uttered simultaneously with use-actions, require instructions for subjects to do so. Any instruction turns a user into a subject and a natural setting into a research context, whatever the encouragements to behave normally. Any evidence of a research context, such as overt observational intentions, tends to evoke an instructional element, if only the encouragement 'to take no notice' and to go about as usual. Thus, the experiencing of a research context and being instructed seem to be different sides of the same coin in terms of the possible consequences of obtrusiveness.

Often, a research context simply cannot be concealed, even if no 'thinking aloud' is applied. Then, the ideal of unobtrusiveness may be approximated, if acceptable from an ethical point of view, by deliberately putting subjects on the wrong track, such as by creating a diversive research context or by misleading tasks. Hence, obtrusiveness may come in varying degrees. See in this respect the study into the use of different coffee creamer cups by Kanis and Wendel (1990). In this study, subjects were first seated in an obvious research environment, i.e. with cameras and equipment clearly visible. Then, each subject was offered a cup of coffee before being informed about the research topic. In this way, half of the subjects (n=17) used their first cup spontaneously, in the sense of not being instructed. On average, these subjects clearly needed more attempts to open a newly designed cup than the informed subjects. This is not to suggest that any perceived research context will always have consequences, see Weick (1968) who reports that subjects, after some initial attention paid to recording devices, tend to lose interest. However, the less obtrusive a research context, the less discussion will be needed in order to demonstrate the impact of observations under certain circumstances on other circumstances, see Figure 4.2. This concerns the issue of generalisability.

4.5   GENERALISABILITY OF OBSERVATIONS

Generalisability involves the extent of invariance of observations for differences in the parameters presented in Figure 4.2, i.e. in going from laboratory conditions to a natural environment, from a model to a functioning product, from tasked subjects to users with immanent goals and who do not consider themselves as being observed. This definition of generalisability is based on the concept of invariance, and does not cover the possibility that it is known how findings vary with differences in circumstances, which would render such findings perfectly generalisable. However, it seems illusory seriously to account for a possible identification of that variance, hence the recommendation above to keep on the safe side by reducing obtrusiveness as much as possible. In addition, if that variance were to be identified as to its overall effects, this would presume a deep insight into human-product interaction, probably rendering the whole enterprise of observational studies redundant.

Questioning generalisations

Numerous combinations of the parameters presented in Figure 4.2 are conceivable: conditions, models, instructions, different types of subjects. This variety is further increased by the diversity in techniques. This includes the choice of thinking aloud with the possibility of involving more than one subject simultaneously in order to stimulate the verbalisation of thoughts cooperatively; likewise the choice of retrospective revelation, which may be enhanced by immediately confronting subjects with the recorded observations, including concurrent thinking aloud. Note that these different techniques are necessarily obtrusive, if only momentarily.

It is not the issue that any research context, in creating its own specific artifacts, would always be the source of dramatically biased predictions if findings are generalised, without any reserve, to the ideal of Figure 4.1. The point is that bias cannot be excluded in such generalisations, while at best some global notions are available about a probable direction of possible deviations, such as those due to paying extra attention (see the coffee creamer example above) and the shaping of subjects' reactions by social desirability. However, no matter how convincing the occurrence of possibly biased generalisations may be, during a design process it seems impossible to generate undisputed evidence in order to check any variance of findings based on a model. A model may still anticipate many possible products as 'future realities', see the remark above on simulation.

Even in the case of 'ideal' observation (see Figure 4.1), generalisations may be questionable. See in this respect the study by Loopik *et al.* (1994) who observed subjects in their first confrontation with a newly designed vacuum cleaner and after having used this for some time. The following types of difficulties were distinguished:

- Operational difficulties (*od's*) which turned out to be transient, i.e. inital problems solvable by trial and error, without outside help.
- Quasi-persistent *od's*, i.e. initial problems to be overcome by consulting the operational manual one or more times, e.g. the meaning of a functionality that was encountered for the first time such as an electronic suction power regulation.
- Persistent *od's*, i.e. problems which emerged in the first confrontation and still prevented subjects from smooth operation after a period of usage (operation manual still needed), in particular more advanced technological intricacies like the automatic adjustment of the power suction to the type of floor covering, and a stand-by function.
- *Od's* emerging during actual usage, such as the unintentional operation of a mechanical suction power regulation (in that study the largest category *od's*, generally of a permanent nature.)

These findings illustrate the possible transience of initial difficulties of humans as adapting, learning creatures, as well as the possible difficulty of comprehending the operational difficulties observed in first time usage, both of which set limits to the generalisability of findings.

4.6 DESIGN RELEVANCE IN QUALITATIVE AND QUANTITATIVE RESEARCH INTO USER ACTIVITIES

Design relevance versus measurability

Above, a right-to-left increase of design relevance is demonstrated of human involved data within the user block. In Kanis (1998) it is argued that this increase in design relevance goes along with a right-to-left decrease in measurability/observability. This decrease becomes manifest in the limited or even non-existent possibility to claim repetition and to establish deviation, particularly in the case of self-reports (Kanis 1998a). In addition, observation of use-actions tends to be cumbersome. For that matter, these difficulties in measuring and observing user activities, including any effort involved, cannot be identified as particular consequences of a quantitative or a qualitative approach.

Quantitative versus qualitative usage centred research

In as far as these approaches can be clearly distinguished, a key-issue in quantitative research is to achieve comparability between observations by scaling them on a particular dimension such as time, or by lumping observations together in a limited number of categories. Then, measurement results are achieved by establishing the frequency with which adopted units of measurement/observation occur. As a rule, information is lost in this process, e.g. sacrificed to averages.

Generally, qualitative research is geared to the identification of meanings. There are no fixed units of measurement/observation laid down *a priori*, as differentiation must be explicitly allowed for in accordance with findings, with the extreme being that each observation turns out to be unique.

Table 4.1 suggests that the design relevance of the results of research into user activities requires a qualitative approach. However, this need not necessarily be the case. Phenomena which can be described quantitatively in a dimension that is directly usage-relevant, should not be excluded from being studied quantitatively. For instance, if the question of what movements people make with their feet in driving a car has to be answered, preferably these movements should be recorded in time and place, the latter, if possible, three-dimensionally. If the question is whether people tend to push a button long enough in order to activate a particular functionality, then not to measure time may be a serious shortcoming. Note the difference with time as an additional dimension resulting from the transformation of operational difficulties into time consumption, which may be completely insignificant for users involved. Overall, it is true that the aim of usage centred research for everyday product design, i.e. finding out what users actually do, how and why, should primarily be accommodated with a qualitative approach.

The scientific status of qualitative usage centred research

In promulgating quantitative research ('hard science') as superior to qualitative research, social scientists frequently cite the following quote by Lord Kelvin: "When you cannot express it in numbers, your knowledge is of a meagre and unsatisfactory kind". In some circles, scholars never cease to emphasize that quantification is only possible after thorough scrutiny of how to count and in what categories. When the latter is known, counting may just be superfluous. Labov (1972: 258) says: "Quantitative research implies that one knows what to count, and this knowledge is reached only through a long period of trial and approximation, and upon the basis of a solid body of theoretical constructs. By the time the analyst knows what to count, the problem is practically solved." Here it should be noted that Kelvin's assertion (see above) is cited out of context, as the complete quote runs as follows (Sydenham 1979: 4): "In physical science [...] when you can measure what you are speaking about, and express it in numbers, you know something about it; but [...] when you cannot express it in numbers, your knowledge is of a meagre and unsatisfactory kind: it may be the beginning of knowledge, but you have scarcely, in your thoughts, advanced to the stage of science, whatever the matter may be." So, to rephrase Kelvin, it may be observed that in everyday product design, when you can only express what you are speaking about in numbers, your knowledge is of a meagre and unsatisfactory kind.

How supportable is small sampling?

The last issue to be raised in this context is the number of people to be involved. Compared to qualitative research, quantification tends to be carried out with large numbers of observations, both in order to arrive at estimates with sufficiently narrow confidence intervals and because qualitative research is time consuming. What does this imply in terms of the number of subjects to be involved during a design process with tight time (and money) constraints? The impact of the number of observations can be clarified on the basis of the formula:

$$P_+ = 1-(1-p)^n$$

with p as the probability to come across an event, let's say a particular *od*, and
P_+ the probability to observe that *od* at least once after n subjects.
Although p is unknown, and, consequently, P_+ cannot be established, the formula shows
that, the lower n,

(i) the more improbable it is that an observed *od* has a low p, and
(ii) the more probable it is to miss an *od* with a relatively high p.

The latter point, (ii), constitutes the weakness of small sampling: not observing what is
not rare. The first point, (i), constitutes the strength of small sampling: what is observed
has a low probability of being rare, see the illustration in Figure 4.3.

Figure 4.3 The low probability of small sampled observations to be rare.

Usage centred research in a design context is primarily geared to distinguishing
significant cases, rather than lumping cases together in categories (see above). Thus,
ideally, observation should be stopped as soon as the probability of coming across
something new with the next subject lies below a certain minimum. Seemingly, usage
centred research can only approach this ideal in a very general way. A practical reason is
that for estimates with sufficiently narrow confidence intervals, too many subjects would
be needed to allow for the constraints of design contexts (see above). A more
fundamental reason is that the monitoring of any user trialling on the basis of the
computation of those estimates presupposes that it is clear in different phases of the
trialling which categories are worth distinguishing. The obvious problem is that this
insight tends to be settled afterwards, as this type of research is bound always to be
exploratory: if it were not, the whole enterprise would be a complete waste of time and
money, see Labov (1972).

4.7   REFERENCES

Gelderblom, G.J., 1998. *User Cognition in Product Operation (thesis)*. Faculty of Industrial Design Engineering, Delft University of Technology.

Green, W.S., Kanis, H. and Vermeeren, A.P.O.S., 1997. Tuning the design of everyday products to cognitive and physical activities of users. In *Contemporary Ergonomics*, S.A. Robertson (Ed.), Taylor & Francis, London, pp. 175-180.

Kanis, H., 1998. Usage centred research for everyday product design. *Applied Ergonomics*, **29**, 1, pp. 75-82.

Kanis, H., 1998a. Design relevance of usage centred studies at odds with their scientific status? In *Contemporary Ergonomics*, M.A. Hanson (Ed.), Taylor & Francis, London, pp. 577-581.

Kanis, H. and Van Hees, L.W., 1995. Manipulation of pushbuttons and round rotary controls. In *Proceedings Human Factors Society 39th Annual Meeting*, pp. 374-378.

Kanis, H. and Wendel, I.E.M., 1990. Redesigned use: a designer's dilemma. *Ergonomics*, **32**, pp. 459-464.

Labov, W., 1972. *Sociolinguistic Patterns*. University of Pennsylvania Press, Philadelphia.

Loopik, W.E.C., Kanis H. and Marinissen, A.H., 1994. The operation of new vacuum cleaners, a users' trial. In *Contemporary Ergonomics*, Robertson, S.A. (Ed.) Taylor & Francis, London, pp. 34-39.

Rasmussen, J., 1986. *Information Processing and Human Machine Interaction: An Approach to Cognitive Engineering*. Amsterdam, North Holland.

Rooden, M.J., 1998. Thinking about thinking aloud. In *Contemporary Ergonomics*, M. Hanson (Ed.), Taylor & Francis, London, pp. 328-332.

Sydenham. P.H., 1979. *Measuring Instruments: Tools of Knowledge and Control*. Peter Peregrinus Ltd, Stevenage.

Vermeeren, A.P.O.S., 1997. Instructions in user trialling: setting tasks or describing contexts. In *Proceedings of 13th IEA-congress*, **2**, Finnish Institute of Occupational Health, Helsinki, pp. 177-179.

Weick, K.E., 1968. Systematic observational methods. In *The Handbook of Social Psychology*, G. Linsey and E. Aronson (Eds), **2**, Research Methods, Addison Wesley, Reading, pp. 357-451.

CHAPTER FIVE

Designing Scenarios and Tasks for User Trials of Home Electronic Devices

ARNOLD P.O.S. VERMEEREN

Faculty of Industrial Design Engineering, Delft University of Technology,

Department of Product and Systems Ergonomics,

Jaffalaan 9, 2628 BX Delft, The Netherlands

5.1 INTRODUCTION

Essentially, user trials are simulations of product usage, in which subjects (who are not actual users) try to fulfil tasks (usually imposed by an experimenter) in an experimental setting by using a product or product simulation (e.g. prototype, mockup, sketches) (see Kanis and Vermeeren 1996). In a seminar discussion on usability evaluation in industry (Jordan, Thomas and McClelland 1996) the artificiality of laboratory environments and evaluation protocols was discussed. It was stated that: "Laboratories will often be free of the sorts of distractions that users will encounter when using a product in its real context of use and (that) evaluation scenarios often request users to perform tasks in isolation which, in reality, they might do whilst also undertaking other activities...". There was discussion of how scenarios could be used in order to find a balance between, on the one hand, control over the evaluation and, on the other hand, generating outcomes that say something about real-life product use. It was stated that: "It is (also) possible to generate a scenario through the way in which the tasks are set. For example, instead of simply setting users a sequence of tasks to do, it may be possible to develop a 'story', perhaps by asking participants to pretend that they are, say, secretaries working for a particular type of company and presenting the tasks as if they were problems that might arise during the working day". However, further on in the discussion, it was stated that: "... it was considered to be highly questionable as to whether ecological validity can really be achieved this way". In this chapter we have a closer look at this problem. We focus on the problem of designing tasks for user trials.

Not much can be found in the literature on selecting and formulating tasks for user trialling. In their book *'The Practical Guide to Usability Testing'* Dumas and Redish (1993) devote two chapters to tasks and scenarios. These chapters give an elaborate picture of the process of selecting and formulating tasks and scenarios. However, the potential methodological consequences of the selection of tasks is not discussed. Dumas and Redish distinguish tasks from scenarios and advocate the use of scenarios in user tests: "scenarios describe the tasks in a way that takes some of the artificiality out of the test". For example, the task 'save a number in the telephone's memory', becomes a scenario when it is written as:

You can save numbers in the telephone's memory and then call those numbers without dialling all the digits each time.

Your best friend's number is 212-555-1234.

Put your friend's number in the telephone's memory.

However, it remains questionable whether this different way of formulating the task, really takes much of the artificiality out of the trial. In general, product use is considered to be driven by goals. Tasks in user trials can be seen as substitutes for spontaneous user goals. In real-life, users may set themselves goals, or goals may be imposed by, for example, their work. In user trials the goals are provided by the experimenter (in the form of tasks). It is crucial that the selected tasks are realistic; unrealistic tasks provide subjects with unrealistic goals which in turn may lead to unrealistic behaviour and thereby to outcomes that are not predictive for problems in actual product use. In turn, this may lead to wrong design decisions (see above, the discussion about artificiality in user trialling).

From this point on, we focus on user trials that are meant to investigate the problems that users may have in learning how to use home electronic devices by exploration. Draper and Barton (1993) found that during such learning processes, users may set themselves two types of goals: experimenting with commands to learn what they do (we will call that 'learning' goals), and looking for commands to achieve specific effects ('product setting' goals). Usually, in user trials subjects are asked to make specific settings on a product. Thus, subjects are provided with 'product-setting' goals. In addition, subjects are often asked to stick to the tasks and perform them as accurately or as fast as possible. Thus, they are not stimulated to set themselves 'learning' goals and to deviate from the 'product setting' tasks. In practice, however, it could be that users do set themselves 'learning goals'. Therefore, theoretically, conducting user trials in the way that is described above might lead to outcomes that are not generalisable with respect to explorative learning processes in practice. If user trials are conducted in a design context this could mean that wrong design decisions will be taken on the basis of these outcomes.

Below we describe a study in which we investigated whether asking subjects to perform 'product setting' tasks, leads to different user trial outcomes as compared to when subjects are not provided with such tasks. Based on our findings we point at some factors to consider in designing tasks for user trials.

5.2 METHOD OF THE STUDY

Comparative user trials were devised with 10 subjects in a between-subject design. As the object of evaluation a programmable thermostat (Honeywell Chronoterm III, see Figure 5.1) was used. Subjects were divided into two equally sized groups of five and were each assigned to either a T(task)- or an E(exploratory)-condition. In the T-condition subjects were given 12 tasks on paper. They were told that it did not matter whether they finished all 12 tasks, but were asked to try and finish each task successfully. However, they were told they could skip a task if after some time they thought they would not succeed. Tasks were formulated as scenarios, in terms of user goals that describe a desired behaviour of the heating system. For example: "You are going away for summer holidays and don't find it necessary that the house is heated during that time. Set the thermostat so that the house is not heated during the holidays". In the E-condition subjects were asked to spend 25 minutes on getting acquainted with the thermostat and learning how to use it. They were told to imagine they had moved into a new house where this thermostat hangs on the wall and that they now wanted to find out how the thermostat worked. They were told that it did not matter whether they succeeded in making any realistic settings in the end, or whether they completely understood the way of programming the thermostat. As the thermostat was not connected to a real heating system, there was no real-life feedback. The thermostat in itself was completely functional and even indicated when it 'tried' to switch on the heating. None of the subjects had any previous experience with using a programmable thermostat, but the subject groups were comparable with regard to the experience they had in simple programming of home electronic devices (like VCRs and CD-players) and in computer use.

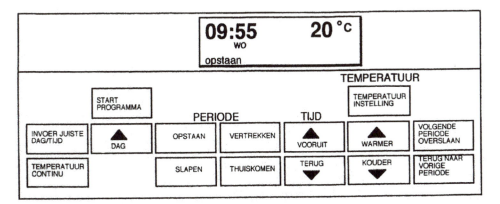

Figure 5.1 A schematic representation of the Honeywell Chronotherm III. Buttons are grey except for 'warmer' (red) and 'kouder' (blue). All labels are in Dutch.

All sessions were recorded on video (both subjects' hands and an overall picture were videotaped). After each session, the just-recorded video tape was discussed with the subject; subjects were questioned about all instances in the session where they seemed to have some kind of operational problem, or where the reason for a button press was unclear. Subjects were repeatedly reminded only to verbalise the reasons for their behaviour if they really remembered them. Complete protocols of button presses and verbalisations were constructed, forming the basis for inferring user goals. Subsequently, lists of user problems were derived from the combination of user goals and protocols. All instances in which subjects were not able to achieve a goal or where this took extreme effort were regarded as problems. The user goals, as well as the user problems and the utterances and observations were analysed to answer the research question at hand.

5.3 RESULTS

The results are presented in Table 5.1. Note that, in the T-condition, the tasks were formulated in terms of user goals. However, in some cases subjects did not succeed in going through all the tasks because of lack of time. Moreover, some subjects spontaneously set themselves goals. Therefore, the number of user goals does not equal the number of tasks provided by the experimenter.

If we compare the user goals as well as the observed user problems in the two conditions, the following results show up:
(a) In both conditions product-setting goals were set (either through the given tasks or by the subjects themselves). However, this was only true for those product-setting goals that related to the most basic functionality of the product (user goals 1, 2, 3 and 14).

(b) In trying to achieve user goal 3 two subjects in the E-condition set a parameter to an invalid setting and did not notice this. A problem like this could never have been found in the T-condition, as in that case parameter values were given in the tasks.

(c) Product-setting goals relating to non-standard, or less frequently occurring situations of use were only set in the T-condition (user goals 4, 9, 10, 12, 13, 15 and 17; in most of these cases users had to overrule the just entered 'standard' daily program to achieve the desired settings). Examples: settings for having a day off, going for a holiday, etc.

(d) Because user *goal* 4 was only set in the T-condition, user *problem* 4 showed up only in that condition as well. As four out of five subjects experienced this problem, it is clear that here the E-condition missed something.

(e) User problem 17 is a special case. Here subjects were provided with a goal (17, 'skip the period after the next') that was somewhat similar to user goal 9 ('skip the next period'). However, contrary to user goal 9, there was no special function available to achieve it. It had to be done by changing the weekly program, a task that every subject was able to do. Nevertheless, three out of four subjects did not manage to achieve user goal 17. They kept searching for a special function to achieve the goal, and apparently refused to believe that it had to be done by changing the weekly program. A possible explanation for this could be that, due to the fact that the experimenter imposed the goals upon them and the fact that the task looked a bit like user goal 9 the subjects believed that there had to be some special function ("there must be something, because they asked me to do this...").

(f) Product-setting goal 19 ('switch off the thermostat') also was only set in the T-condition. However, this was not a goal imposed by the given tasks. It was an unachievable self-imposed user goal resulting from an erroneous interpretation of how to execute the task, triggered by its wording (i.e. "At 11 p.m. you go to bed, the heating may be off then"). This formulation made subjects try to turn off the thermostat, whereas in fact the thermostat had to be set to a low temperature so that the *heating* would automatically switch off; not the *thermostat*).

(g) In the E-condition some subjects set product-setting goals that were unachievable because they relied on some conceptual misunderstanding of the general procedure for programming the thermostat (user goals 20 and 21).

(h) Learning goals almost only showed up in the E-condition (user goals 7, 8, 11, 16, 18).

(i) User goal 5 ('verify programmed settings for weekdays') indicates that subjects in the E-condition felt a need to verify the settings they had made, most likely as part of achieving a learning goal. In the T-condition subjects usually only did so if they were explicitly asked to look up settings (as was the case with user goal 6).

(j) User problems 7 (occurring while trying to achive learning goal 7, 'learn what the button *volgende periode overslaan* does') and 9 (in trying to achieve product-setting goal 9, 'skip the next period; i.e. use *volgende periode overslaan*') form a special case. Although they are different in nature and relate to different user goals, they do relate to the same product function. So, from both conditions a designer could conclude that something is wrong with the design of that function. However, they give slightly different indications as to the causes of the observed problems. User problem 7 suggests that it is hard to read from the label what the button is for, and that the feedback after using the button does not convey the function properly either. User problem 9 only suggests that the feedback after using the button does not convey the status of the product adequately.

(k) A similar difference occurs with user problems 11 and 12. User problem 11 tells us that, from the label and the feedback subjects can not read what the button is for (i.e.

that function there is), whereas in the case of user problem 12 the subject does not succeed in using the button (due to status feedback problems).

Table 5.1. User goals and major user problems as found in the experiment. Figures in the columns are numbers (#) of subjects (Ss).

User goals (self-imposed or set by the tasks)	# of Ss E-cond.	# of Ss T-cond.	Major user problems	# of Ss E-cond.	# of Ss T-cond.
1. Set current day and time	5	5			
2. Programming weekdays	5	5			
3. Programming one or both days of the weekend	5	5	3a. Subject forgot to program one day	-	1
			3b. Subject sets starting time for last period at or after 12 p.m.	2	-
4. Delete program for the weekend	-	4	4. No success	-	4
5. Verify programmed settings for weekdays	5	-			
6. Verify programmed settings for the weekend	1	5			
7. Learn what 'volgende periode overslaan' is for	5	-	7. Subject does not find out what it is for	4	-
8. Systematically verify whether 'volg. periode overslaan' is well understood	3	1			
9. Skip the next period (i.e. use 'volgende periode overslaan')	-	5	9a. Subject seriously doubts success	-	1
			9b. No success, but subject does not notice	-	2
10.Undo 'skip the next period'	-	5			
11.Learn what 'temperatuur continu' is for	5	-	11. Subject does not find out what it is for	5	-
12.Set temperatures for the holidays (i.e. use 'temperatuur continu')	-	3	12. No success	-	1
13.Undo setting for the holidays	-	3			
14.Set temporary higher or lower temperature	3	5			
15.Undo temporary settings for the current period	-	5			
16.Learn what 'terug naar vorige periode' is for	4	-			
17.Skip the period after the next	-	4	17. No success, but subject does not notice	-	3
18.Learn what 'temperatuur instelling' is for	3	1			
19.Switch off the thermostat	-	4			
20.Fix a temperature setting	2	-			
21.Programming one single weekday	2	-			

5.4 CONCLUSIONS

If we look at the studied problems we see that tasks and scenarios in user trials can affect subjects' behaviour and observed user problems in five different ways, namely through:

1. the type of task,
2. the wording of the task,
3. the parameters and parameter values embedded in the tasks,
4. the sole fact that subjects are provided with tasks, and
5. the order in which tasks are provided.

Type of task

In the T-condition tasks were formulated in terms of goals for a product status to be achieved. In the E-condition the instruction was more or less to explore the interface. The results shows that the T-condition made subjects try to achieve 'product-setting goals' (see (a)), whereas the E-condition made them try to achieve both 'product-setting goals' and 'learning goals' (see (a) and (h)). This results in the user trying to perform different types of tasks. To illustrate this we can use the findings of Draper (1993). Draper states that in task analysis often real tasks faced by users are missed. He states that: "For each command the designer is often only thinking of supporting one 'task', yet up to four are in fact commonly at issue":

1. performing the function,
2. verifying success,
3. discovering how to perform the function, and
4. discover the function of commands (given their visible presence).

If we translate this to product-setting goals, we see that the minimum tasks needed to perform them are task types 1 and 3 ('performing the function' and 'discovering how to perform the function'. If users perform only these tasks, they can be succesful in achieving the product setting goal. Task type 4 ('discover the function of a command') directly relates to 'learning goals' (by definition). Results (a) and (h) show that task type 4 hardly ever occurs when subjects are provided with the type of tasks we provided them with in the T-condition. Result (i) shows that task type 2 ('verification of success') happens more often in the E-condition than in the T-condition. In the E-condition subjects often spend considerable time in trying to verify the actual status of the product after having used a command. Obviously, because they were often trying to learn something, and had to rely on status feedback in order to progress. In the T-condition, there was a lesser need for verification. If subjects were not sure about the success of a certain action, they always had the choice to 'decide' that they had enough confidence in the result and then proceed with the next task. Thus, the results suggest that if an evaluator is interested in task types 2 and 4, explicit action has to be undertaken, other than just asking subjects to try and achieve a certain product status. Either, one should try to get subjects into a 'learning' mode, or tasks should be formulated in such a way that users spontaneously set themselves goals in accordance with tasks types 2 and 4.

Results (j) and (k) show that as a consequence of the fact that subjects set themselves different goals in the different conditions, the observed user problems and their likely causes may also be different, even though they may be related to the same product functions. As a consequence, if the outcomes are used as input for redesigning the product, the design decisions meant to fix the observed problems may also be different.

The wording of the task

Result (f) shows that the specific wording of a task may be of crucial importance to the user problems observed in a user trial. Although the text used in the task was literally correct, it still suggested a different meaning to users. Another example from the study (which did not result in a user problem) is where wording suggested how to perform a task: in one task subjects were told they went on a holiday and that the heating should be off then. Some subjects (unsuccessfully) sought the solution for that task in (temporarily) changing the thermostat's program. In the subsequent task, subjects were told to imagine they came back from holiday and were asked to set the thermostat such that the regular program functioned again. One subject who initially had not succeeded in making the holiday settings, realised how he should have done it, after reading the subsequent task: it made him realise that he should not have tried to change the program temporarily, but that he should have tried to bypass it. Although here this did not result in extra user problems, or in 'missing' user problems, it showed how subtle the influence of the specific wording of a task can be on the behaviour of the subject.

The parameters and parameter values embedded in the tasks

Results (b) and (g) provide indications for the importance of considering particularly what parameters and parameter values to mention in the tasks. Result (b) reports a user problem of a subject who sets an invalid parameter value. However, to the user the value is very normal. This makes clear that once parameter values are mentioned in the task, this deprives the evaluator of the possibility of finding out what the values are that users would want to set or consider as 'normal'. Result (g) shows another way in which mentioning parameters and values can influence subjects' behaviour. In the E-condition no parameters were mentioned whatsoever. User goals 20 and 21 were based on a conceptual misunderstanding of what parameters can be used in programming the thermostat and how, i.e. on the structure underlying the way the thermostat is programmed. These problems occurred in the initial exploration phases. These misunderstandings did not occur in the T-condition, presumably because there all parameter values to be set were given in the tasks, thereby revealing the parameters the program makes use of (i.e. working days as one block, four periods in a day, characterised by their starting times, and with independent temperature values). Mentioning those values in the tasks, even though disguised in a story, apparently triggers ideas for how the user thinks the program is structured.

The sole fact that tasks are given

Result (c) reports the obvious finding that with tasks one can make people use functions that they otherwise would not use. The type of functions mentioned in result (c) (i.e. non-standard functions, or functions needed in less frequently occurring situations) are not applied spontaneously by subjects if they are not asked to do so. On the one hand because they may not think of situations in which those functions can be useful, and on the other also because a product does not always clearly convey what the function buttons can be used for (i.e. they do not immediately trigger a function to the user). Result (c) is one of the very reasons why tasks usually are applied in the first place, one can steer the behaviour of the subjects with it. On the other hand, it also shows that if evaluators want to find out whether subjects can (by themselves) discover what functions a product incorporates, subjects should not be provided with tasks that ask for specific settings for which the use of those functions is required. Result (e) supports this. One of the explanations of this finding is that it seems to indicate that if subjects are asked in a task, to make some non-standard setting, they readily assume there must be some special function for that. Another explanation for this latter finding has to do with the order in which the tasks are provided.

The order in which tasks are provided

Result (e) can also be explained by the fact that the task resembled a (not immediately) preceding task. The preceding task had to be solved by bypassing the settings of the period after the current period (this could be achieved by pressing the button 'volgende periode overslaan' ('skip next period'). In this task subjects were told that the settings for the period after the next had to be different from those in the regular progam. For that task there was no special button, it had to be done by changing the week program. The preceding task could have made the subjects think that the same button should be used for that, as it is known that people tend to apply known procedures for new situations if those situations resemble familiar situations (e.g. see Rasmussen 1986).

5.5 SUMMARY

Summarizing, we can conclude that designers of tasks (or scenarios) for user trials have to consider a number of aspects that relate to the value and type of outcomes from the trial. The designer should be aware that:

- A real-life task does not only consist of 'discovering how to perform a function' and of 'performing that function', but that 'verifying success' is often an important part of the task as well. Also, users sometimes apply a command just to find out what it does. The latter two tasks do not always come naturally if subjects are asked to perform 'product-setting' tasks. Therefore, 'product-setting' tasks may not be sufficient to study how well future users are able to read from the product in what state it is and what the effect of a command will be. Explicit tasks for verification of success or for finding out the function behind a command may be needed for that.
- Each and every word in the task description can trigger a specific way of thinking (either correct or incorrect) about how to use the product. It is therefore of utmost importance that every word is carefully chosen (no improvisation by the experimenter) and if possible tried out in a pilot trial.
- If part of the product use consists of setting parameters there are two things to consider:
    - if the parameter values to be set are given in the task, it is no longer possible to find out what values subjects would naturally choose;
    - sets of parameters that are given in a task (either with or without values to be set), can provide the subject with hints as to how the procedure of programming the product is structured.
- Tasks make subjects search for, and apply specific functionality; the advantage of that is that one can be reasonably sure that problems in the use of that functionality

will be revealed; a disadvantage may be that it is no longer possible to find out whether users can by themselves find out that such functionality exists at all.

- The order in which the tasks are given is important and should therefore be carefully chosen. Usually, subjects in a user trial have only very limited knowledge and experience directly related to the use of exactly the product that is trialled. Therefore, knowledge and experience gained in early (usually relatively simple) tasks is very likely to be applied in subsequent (more complex) tasks and will determine to a large extent how well those tasks will be performed.

5.6 REFERENCES

Draper, S.W., 1993. The notion of task in HCI. In *Proceedings of InterCHI'93 Adjunct*, ACM, New York, pp. 207-208.

Draper, S.W. and Barton, S.B., 1993. Learning by exploration, and affordance bugs. In *Proceedings of InterCHI'93 Adjunct*, ACM, N.Y., pp. 75-76.

Dumas, J.S. and Redish, J.C., 1993. *A Practical Guide to Usability Testing*. Norwood, Ablex, NJ.

Jordan, P.W., Thomas, B. and McClelland I.L., 1996. Issues for usability evaluation in industry: seminar discussions. In *Usability Evaluation in Industry*, P.W. Jordan *et al.* (Eds), Taylor & Francis, London.

Kanis, H. and Vermeeren, A.P.O.S., 1996. Teaching user involved design in the Delft curriculum. In *Contemporary Ergonomics*, S.A. Robertson (Ed.), Taylor & Francis, London, pp. 98-103.

Rasmussen, J., 1986. *Information Processing and Human Machine Interaction: An Approach to Cognitive Engineering*. North Holland Publishing, Amsterdam.

CHAPTER SIX

Implications for Using Intelligence in Consumer Products

JOHN V. H. BONNER

Institute of Design, Teesside University, Middlesbrough,

TS1 3BA, UK

6.1   INTRODUCTION

This chapter explores the implications for the use of so-called 'intelligence' in consumer products by drawing upon research work that has been undertaken within the field of Human Computer Interaction (HCI). By taking a non-specialist perspective of the field of artificial intelligence and reviewing the emergence of intelligence from a human factors and product design approach, issues are raised concerning intelligent or 'smart' products. In this chapter, the following question is considered: is the use of intelligence inevitable or does the concept of products capable of thinking reach beyond the normal boundaries of acceptance behaviour for new technologies?

Historically consumer products with an interactive user interface have been too diverse in functionality and too lacking in complexity to require intelligence. However, the pervasive use of the micro-processor has resulted in consumer products having many more features for the user to learn. Also, products now use electronically rather than mechanically based control and display devices where feedback to the user often has to be designed into the interaction process. Therefore, under these circumstances, the notion of product intelligence is being considered as a way of supporting communication between the user and the product. Currently, very few consumer products use any form of artificial intelligence, although some products use simple heuristic rules, algorithms, or fuzzy logic for function control or display. For example, some photocopiers will automatically select the correct paper size corresponding to the size of the paper on the photocopying bed. Some of the obvious benefits of using intelligence in user/product interaction could be:

- to support and accommodate user variation and needs,
- to support and accommodate different types of tasks,
- to allow users to use more complex devices without extensive training.

Focusing on the final point above, many traditional products such as microwaves and video cassette recorders (VCR) may become more complex with new added features. However, there is also evidence that a new breed of consumer products is emerging which could be far more intimidating and difficult for the consumer to operate. These are information based products such as 'Web TV' which allows access to the Internet using the television or interactive TV. Consumers will have access to huge information resources and have greater control over how they use their personal 'technological environment'. Many manufacturers may consider using intelligence to help navigate and support the user through these information resources. Further evidence can be found in the way that information is processed. Products are shifting from analogue to digital processing which has implications for how we will potentially work and communicate in the future. Negroponte (1995) and Shneiderman (1990) provide some predictive reviews of how technology will alter personal, domestic and social interaction using these information based products.

More immediately, however, there are technological developments making information based services available today where the use of intelligence could be considered, such as:

- Increased telecommunication transmission rates (Cochrane and Smith 1994) which allow access to high 'bandwidth' data, such as navigating through 3D models in real time.
- The convergence of telecommunications, computers and consumer products to produce more information based products such as access to the internet using conventional television, video-on-demand, and creating personalised channels on digital television (Netravali and Lippman 1995).
- Personal information and communication technologies (Zimmerman 1995), for example personal digital assistants (PDA), mobile telecommunications, pagers and even wearable technology (Bedford 1997).

In these examples intelligence could be used to adapt and fit the selection and flow of information to user needs. Supporting this, control and display technology is also advancing with many large Japanese electronics companies including Sony, Sharp, Toshiba and Panasonic developing ultra-thin flat screen systems (Cole 1997) and producing high resolution touch displays which may increase the level and types of information displays to a wider range of consumer products.

Some intelligent products are currently available such as the Sharp Logicook Microwave (Nomura & Brownlow, 1996) or are under development, for instance adding intelligence to a Home Electronic System (Schoeffel 1995) manufactured by Bosch Siemens Hausgeräte. Other candidate examples for intelligent products include lifts, packaging, prosthetics, paint and warning devices (Fleck *et al.* 1997).

6.2   DEFINING PRODUCT INTELLIGENCE

In order to address the issue of using intelligence, there must first be some understanding or interpretation of the concept. In other words, what is meant by an intelligent or smart product? There is, to date, little research attempting to define or categorise levels or types of intelligence in consumer products. Although considerable work in using intelligence in computing has already been undertaken in the field of Human Computer Interaction (HCI), Chignell and Hancock (1988) have defined an intelligent interface, as opposed to the concept of intelligence itself, as being:

"An intelligent entity mediating between two or more interacting agents who possess an incomplete understanding of each others' knowledge and/or form of communication".

Frameworks for understanding the salient issues and components of an intelligent interface (Benyon and Murray 1993) have also been presented where an interface, to be able to adapt or exhibit intelligence, requires the system to have three models: a model of the user; a domain model containing a representation of the application; and finally an interaction model between the user and the system where some form of monitoring can take the place of past and current dialogue. Interestingly, even here, the authors express concern about over stretched expectations of using intelligent dialogue in many computer based applications. While a definition and description of an intelligent interface can be found within the field of HCI, little was found that actually attempted to define or describe artificial intelligence itself.

Nevertheless, these broad descriptions of an intelligent interface, as opposed to defining an intelligent interaction, suggest that a complex dialogue between the 'interacting agents' using a set of user, domain and interaction models is required before any real intelligent dialogue can occur. However, the problems of usability and acceptance of intelligence can emerge long before real product intelligence exists. Some VCRs, for example, will automatically select a slow recording speed if the remaining tape cannot accommodate the user's selection. This could be perceived by the user as product intelligence, although the system is complying to simple algorithms. Using the above definition for an intelligent interface, the VCR does not need any understanding of the users incomplete knowledge, only that an operational rule has been violated and that a different function needs to be executed in order to overcome the problem. According to the definition above, the VCR is not intelligent, but the user may perceive the product as intelligent. Little is known about how users might accept product intelligence (real or perceived) although it would appear in using computers that perceived control is an important factor in whether or not technology is adopted by the user (Hill *et al.* 1987). This suggests that intelligence allowing devolved control between the user and product may meet with severe acceptance problems because of this perceived lack of control.

6.3   THE USE OF INTELLIGENCE IN HCI

Within the field of HCI there has been a shift from the concept of artificial intelligence (AI) to the use of autonomous agents which appear to have a future in building intelligent systems. Maes (1996) defines autonomous agents as "computational systems that inhabit some complex, dynamic environment, sense and act autonomously in this environment and, by doing so, realise a set of goals that they are designed for". Autonomous agents can be manifested in different ways such as autonomous robots, two and three dimensional animated agents occupying simulated physical environments, and software agents. The architecture for these agents is highly complex and requires them to be able to perceive their changing environment; internal decisions have to be made in order to achieve the required objective. The agent has to be able to learn and improve its behaviour over time, able to communicate with other artificial and human agents and, finally, be able to translate relevant actions into concrete commands.

Research using 'autonomous agents' in a virtual environment has been investigated by Maes (1996). In one project Maes created an interaction dialogue that allowed a user to interact with animated characters within virtual environments. One of these was a pet dog known as Silas. Users were observed interacting with the dog and their interaction tolerance and behaviour with the dog was noted. It was found, for example, that it was still important to have a human "guide" giving hints on what to do such as telling the subject that patting the dog on the head would change the dog's behaviour. A more tangible example of agent technology is within Internet commerce where agents can be used to search websites against the user's purchasing requirements and then make the purchase on behalf of the user.

Whilst agent technology may be used within consumer products, the nature of an intelligent user/product dialogue could be challenged in its current nature and form. Many consumer products cannot use conventional input devices such as keyboards and mice and may require the exploration of other sensing devices such as field-sensing devices (Zimmerman 1995) which locate and monitor body movement within an enclosed environment. With this in mind, spatial and motion intelligence could be used to enhance an alternative or possibly more natural user/product dialogue, for example, using hand gestures as an input device. Considering intelligence in this way would represent a major departure from conventional HCI dialogues, but again little is known about how these types of 'mediating' dialogue could be used and how viable and acceptable it would be.

6.4   POSSIBLE FUTURE DIRECTIONS

With little known about how intelligence could be applied to consumer products, a certain level of conjecture is required to anticipate or suggest future directions. A good starting point would be to examine different types of interaction dialogue from a broader perspective than any human computer interaction model. Smith (1994) presents a family of five aesthetic languages related to interaction design. By adapting these 'languages', it is possible to evolve them and consider the inclusion of intelligence, for example:

- *Haptic (three-dimensional form and touch)* An example here could be an interface 'surface' which forms and deforms depending on different systems states. In using intelligent polymers (Fletcher 1996) controls could 'appear' when they are required and disappear when they are not required or are unsafe to use. Dynamically changing tactile feedback (control knob resistance to indicate safe and dangerous control positions for example) could also be used as an interaction dialogue to provide enhanced feedback.
- *Auditory (sound and music)* Sound can be a very rich source of information and some research has been undertaken with human computer interaction such as the use of music (Alty 1995) and auditory icons (Gaver 1989). For example, a short sound like a closing door can provide information on whether the room is empty or full, the construction of the door and the force with which the door was closed. An intelligent product could convey complex or subtle messages by changing or modifying auditory displays to enhance the communication process.
- *Literary (prose and dialogue)* This aspect of intelligent interaction has probably received the most attention in the form of voice recognition systems and text-based question and answer dialogues (such as expert systems). Voice recognition systems, while continually improving, still suffer from not offering reliable recognition and an inability to understand a wide vocabulary. Although these methods of dialogue appear to be a natural solution to providing an intelligent dialogue they may appear to be too close in character to human human communication and could then be invested with more capability than is the case.
- *2D and 3D visual displays (such as animation, graphic composition, iconic representation and pictorial space).* This area of interaction is receiving more attention from the research community particularly the representation of complex information, such as databases represented in graphical form, although the concern remains that attention is still required on fundamental visual cues (Diaper and Schithi 1996) such as colour, spatial location and apparent depth. As the use of high resolution flat displays become more ubiquitous, representing and describing intelligent dialogue could be visually based. For example animated icons (Baecker *et*

al. 1991) could be made intelligent by evolving and adapting to changing contextual usage.

These examples of novel interaction dialogues suggest that the scope of intelligent user/product interaction can be broadened beyond the conventional HCI paradigm used today to possibly include less traditional control and displays mechanisms.

6.5   HURDLES TO INTELLIGENT PRODUCTS

An interaction dialogue, however embedded with some form of intelligence, possesses one characteristic distinct from any other type of conventional interaction dialogue. The key difference between an intelligent and non-intelligent dialogue is the implicitly or explicitly devolved knowledge and decision making between the user and the product. This type of user/product interaction is currently extremely unusual and presents an interesting area of study in terms of anticipating the success and acceptance of using an intelligent dialogue.   Predicting the acceptance of any novel or relatively unknown interface technology is difficult although there have been many studies on technology acceptance. Richardson (1987) argues that three major factors need to be addressed to successfully implement technology: usability, utility and user acceptability. Further, research investigating the acceptance of Management Information Systems suggests that managers will use technology, firstly if they perceive it to be a useful tool and secondly, that the technology is fun to use (Igbaria *et al.* 1994). The inference which can be made here is that intelligence, as a functional attribute, must firstly be perceived as providing this utility before any other factors of acceptance will be considered, suggesting that intelligence must contain perceived utility and not be regarded as additional functionality.

Despite this, the challenge for developers of consumer products today is perhaps more pragmatic. With current technology capable of providing actual or perceived smart or intelligent attributes, a more pressing question would be: are users already wary of smart products? Consumers, as users of technology, can exercise a higher level of discretion in both purchasing and using technology as opposed to computer users who, in a commercial context, do not generally make purchasing decisions and often are provided with formal or informal training in new technology. A study by Burford and Baber (1994) illustrates this point very well in the introduction of an adaptive interface for an auto-teller machine. They found that, for the product to be successful, it was important that users were made explicitly aware that they were using an adaptive interface. Such examples demonstrate that the development of intelligent interfaces and an understanding of how users will accept and use them requires considerable attention from the human factors and design communities.

In conclusion, a series of questions and issues are presented which raise some important concerns if intelligence is to be used in consumer products, and which need to be addressed before intelligent products become an integral part of life.

How difficult is it for users to develop mental models of intelligent products? Does the interface need to explain how 'intelligent' it is? Can different levels of intelligence be defined and communicated? Assuming that different levels or types of intelligence emerge, will a product have to explicitly state its intelligent behaviour? For example, a product may have to initially disclose that it learns from the user's interaction behaviour and adapts over a period of time to the user; furthermore, what type and level of feedback should be given to the user to communicate that the product has acted upon some internalised assumptions and that an intelligent decision, as opposed to a predictable action/feedback dialogue, has taken place?

What would an adaptive interface adapt to? Many consumer products have specific functions. If they adapted intelligently, for example, by providing specific functions or features for different users or tasks, there are important implications in the product being fully aware of its contextual use. Also, would the product have to provide some form of rationale before adapting to the user or task? Assuming this is possible, how would the product establish the user's level of competence in order to offer the correct level or type of intelligence or dialogue support? Typically this could be achieved by monitoring and recording user interaction. This is often easier to establish with human computer interaction, where the user may be expected to be involved in more complex tasks over a longer period of time. User/product interaction tasks tend to be shorter and more intermittent making the predictive modelling of user interaction more difficult.

A key design principle is that interfaces should act in a consistent manner. However, once a product becomes 'intelligent' some of this consistency in behaviour could be lost. How would users accept or interpret this? The intelligent dialogue might be interpreted by the user as unpredictable and untrustworthy, particularly if the product is perceived as having made a mistake. What aspects of intelligence will consumers accept or reject in products? Research is needed to investigate what types of devolved decision making are acceptable and effectively understood by the user.

The use of intelligence may allow the user to set instructions in more natural or less precise ways. If this were possible, what would be the most effective ways to support these more 'natural' dialogues and are there any alternatives to speech driven dialogues which have proved difficult to implement widely? Will defacto dialogue styles simply emerge from product manufacturers?

There is also some suggestion from this brief review of artificial intelligence that, beyond understanding and modelling intelligent dialogue, there is another inhibiting factor preventing a prolific development of intelligent products. This is a lack of 'sensing devices' which can provide the product with 'awareness' of the physical and psycho-social context in which the product is operating. The functionality of any 'mediating' device between the user and the product is still constrained by hardware limitations.

Finally, models of intelligent interaction will need to include an understanding of how users may perceive, interpret and understand intelligence. For instance, users may over or under estimate the level of intelligence of the product and therefore make incorrect assumptions about the level of control which the product is capable of, or which is required from the user.

Answers to these types of questions will have to be resolved through extensive longitudinal user trials which go beyond conventional usability measures to ensure that evolving and temporal aspects of any interaction process are fully captured.

6.6 CONCLUSIONS

The questions and issues raised here suggest two distinct hurdles for the successful implementation of intelligence in products. Firstly, the introduction of intelligence must address all three factors of utility, acceptance and usability factors concurrently, all of which run deeper than conventional interaction design problems. Removing the clear delineation between the master and servant raises some important issues about how the decision making process between the product and user is defined.

Secondly, there are development and evaluation hurdles to overcome. Intelligent interaction is not merely a matter of getting the interaction dialogue right it is also about having a much deeper contextual understanding of a broader set of issues, such as have been discussed in this chapter. Jacob *et al.* (1993) suggests that in order to achieve this, interdisciplinary development teams will be required where groups such as electrical and mechanical engineers, anthropologists, ergonomists, software engineers and graphic designers, for example, work together.

This review of current work would suggest that intelligence is not a functional 'add-on'. The possible introduction of intelligent products, if indeed it becomes acceptable to users, will attack many fundamental principles of user/product interaction. The notion of developing thinking machines, however limited, remains an exciting and challenging prospect for designers.

6.7 REFERENCES

Alty, J.L., 1995. Can we use music in computer human communication? In *People and Computers X*, M.A.R. Kirby *et al.* (Eds), Cambridge University Press, Cambridge, pp. 409-423.

Baecker, R., Small, I. and Mander, R., 1991. Bringing icons to life. In *Reaching through Technology, Proceedings of CHI'91*, New Orleans, USA, S.P. Robertson, G.M. Olson and J.S. Olson (Eds), ACM Press, pp. 1-6.

Bedford, M., 1997. Power dressing. *Tomorrow's Technology Today*, **11**, September 1997.

Benyon, D. and Murray, D., 1993. Adaptive systems: from intelligent tutoring to autonomous agents. *Knowledge Based Systems*, **6**, 4, pp. 197–219.

Burford, B.C. and Baber, C., 1994. A user-centred evaluation of a simulated adaptive autoteller. In *Contemporary Ergonomics 1994*, S.A. Robertson (Ed.), Taylor & Francis, London, pp. 46-51.

Chignell, M.H. and Hancock, P.A., 1988. Intelligent interface design. In *Handbook of Human Computer Interaction*, M. Helander (Ed.), Elsevier, North Holland.

Cochrane, P. and Smith, R., 1994. *'21st Century Information Systems.'* British Computer Society, Expert Systems, '94 Conference, pp. 3-17.

Cole, G., 1997. Canon heralds super-flat screen. *Sunday Times*, 2 February.

Diaper, D. and Schithi, P.S., 1996. Red faces over user interfaces: what should colours be used for. In *People and Computers X*, M.A.R. Kirby *et al.* (Eds), Cambridge University Press, Cambridge, pp. 425-435.

Fleck, J., Molina, A. and Nicoll, D., 1997. *Designing Smart Products.* Paper presented to Design Council Research Workshop, London, June/July 1997.

Fletcher, R., 1996. *'Materials and Mechanisms for Force Transduction'.* Physics and Media Group, MIT Media Laboratory (Internal report).

Gaver, W.W., 1989. The SonicFinder: an interface that uses auditory icons. In *Human Computer Interaction*, **4**, pp. 67-94.

Hill, T., Smith, N.D. and Mann, M.F., 1987. Role of efficacy expectations in predicting the decision to use advanced technologies: the case of computers. In *Journal of Applied Psychology*, **2**, 2, pp. 307-313.

Igbaria, M., Schiffman, S.J. and Wieckowski, T.J., 1994. The respective roles of perceived usefulness and perceived fun in the acceptance of microcomputer technology. In *Behaviour and Information Technology*, **13**, 6, pp. 349-361.

Jacob, J.K., Leggett, J.J., Myers, B.A. and Pausch, R., 1993. Interaction styles and input/output devices. In *Behaviour and Information Technology*, **12**, 2, pp. 69-79.

Maes, P., 1996. *Artificial Life meets Entertainment: Lifelike Autonomous Agents.* MIT Website page http://pattie.www.media.mit.edu/people/pattie/CACM-95/alife-cacm95.html.

Negroponte, N., 1995. *Being Digital.* Hodder and Stoughton, London.

Netravali, A. and Lippman, A.B., 1995. Digital television: a perspective. In *Proceedings of the IEEE*, **83**, 6.

Nomura, T. and Brownlow, M., 1996. Thinking microwave oven — Logicook. In *Proceedings of the IEE Artificial Intelligence in Consumer and Domestic Products Colloquim*, 22 October, 1996, Savoy Place, London.

Richardson, S., 1987. Operationalising usability and acceptability: a methodological review. New methods in applied ergonomics. In *Proceedings of the 2nd Iinternational Occupational Ergonomics Symposium*, Zadar, Yugoslavia.

Schoeffel, R., 1995. Sensitive screen HCIs in recent Siemens telecommunication products. In *Symbiosis of Human and Artifact: Future Computing and Design for HCI*, Anzai, Ogawa and Mori (Eds), **1**, Elsevier, Amsterdam, pp. 1167-1172.

Schneiderman, B., 1990. Future directions for human computer interaction. In *International Journal of Human Computer Interaction*, **2**, 1, pp. 73-90.

Smith, C.G., 1994. The art of interaction. In *Interacting with Virtual Environments*, L. Mac Donald and J. Vince (Eds), John Wiley and Sons.

Zimmerman, T.G., 1995. *'Personal Area Networks (PAN): Near-Field Intra-Body Communication'*, **7**, 22 September, Physics and Media Group, MIT Media Laboratory.

CHAPTER SEVEN

Representations of Smart Product Concepts in User Interface Design

SIMO SÄDE

Department of Product and Strategic Design,

University of Art and Design Helsinki,

Hämeentie 135 C, 00560 Helsinki, Finland

7.1   INTRODUCTION

Designing usable and desirable smart products can be supported by early modelling of the product concept, the intended user, the context of use, and the user product interaction. This chapter describes smart products, discusses issues related to their level of usability and likeability, and presents modelling techniques for the early stages of product development. The focus is on the concept development phase, where the most important decisions about the characteristics of the emerging product are made.

7.2   SMART PRODUCTS

Smart products are modern, interactive electronic consumer or professional products. They have original hardware and software, and a specific set of tasks. Their user interface (UI) is limited when compared to computers. Smart products often have a lot of functions, but the high interactivity is managed with minimal resources, typically by menu selection, soft function keys and small displays (Keinonen *et al.* 1996).

Devices we call smart products have been said to be e.g. *interactive consumer products* (McClelland 1987), *intelligent consumer products* (Baber and Stanton 1992), *task-specific digital products with embedded computer technology* or *information appliances* (Carr *et al.* 1994), *high-tech products* or *complex, programmable consumer products, not including computers* (Feldman 1995), and products *with a solid user interface (SUI)* or *products with embedded microprocessors* (Black and Buur 1996).

Today, the term 'smart products' seems to be widely accepted. For instance, the following papers define *products driven by embedded computers* (Caplan 1994), *micro-processor-based, feature-rich products with embedded digital technology* (Bauer and Mead 1995), *modern electronic products* (den Buurman 1997), and *products that contain embedded information technology integrated with a 3D physical product* (Keinonen 1998) to be 'smart products'.

The general trend seems to be that more and more products include information technology and are connected to various information networks. Integration of various products, such as television, telephone, and computer into a multimedia terminal is generally expected to happen. Many portable devices combine the characteristics of computers and smart products, e.g. personal digital assistants (PDAs). Black and Buur (1996) note, that smart product UIs offer advantages over graphical user interfaces (GUIs) in many applications, especially with well defined tasks, in mobile or rugged environment.

7.3 USABILITY AND LIKEABILITY OF SMART PRODUCTS

Usability is *"the extent to which a product can be used by specified users to achieve specified goals with effectiveness, efficiency and satisfaction in a specified context of use"* (ISO 9241 DIS 1994). A product's usability can be measured regarding the learnability, efficiency, memorability, errors, and subjective satisfaction related to its use (Nielsen 1993).

The reasons for the usability problems with smart products (see e.g. Brouwer-Janse 1992; Johnson and van Vianen 1993; Norman 1993; Scott 1993; Stifelman *et al.* 1993; Carr *et al.* 1994; Buur and Windum 1994; Keinonen *et al.* 1996; den Buurman 1997; and Ketola 1997) seem to be related to three issues:

- *The technology-driven design approach and the management of the multidisciplinary process.* In spite of today's user-centered design philosophy, it is obvious that many products are designed in a machine-centered atmosphere. The participants in the product development process are technically oriented, which leads to designs less suitable for non-technical users. The users are required to think in a way which is not natural for them. The designers implement more and more functionality, making the product too complex for the users to access and enjoy the functions, because increasing the number of features by programming is less expensive than it used to be by hardware design. Managing successful seamless design of both hardware and software parts of the UI may be difficult.

- *The character of smart products with their technical limitations.* The physical product and the user interface are disintegrated; the form no longer follows the function. Much of the product is invisible and electronic intelligence takes over some of the control the user used to have. The limited user interface and the vast amount of functions cause problems with understanding the ambiguous, abbreviated language, and with navigation through the complex UI. Another cause for problems are the rigid system constraints — small number of buttons with a small display, limited memory and processing capability, slow data transfer rate, lack of pointing device, the small size of UI items, etc. Testing prototypes in real context is sometimes troublesome, if not impossible.

- *The heterogeneous user groups of these products.* The users of smart products are a heterogeneous group because many smart products are consumer durables. The level of experience, motivation, and skills of these people may vary a lot. Generally, people do not want to spend time learning how to use a product. In addition, these products are often used in difficult circumstances, and as a secondary action for the user at the moment of use.

Usability engineering emphasises the user performance related characteristics of a product. *Likeability* is another essential attribute in smart products. Studying the user's performance alone is inadequate, when considering the user's experience of a product.

The user's feelings are crucial to the formation of the experience. Shackel (1991) defines product's acceptability: "**Utility** (will it do what is needed functionally?), + **usability** (will the users actually work it succesfully?) + **likeability** (will the users feel it is suitable?) must be balanced in a trade-off against **cost** (what are the capital and running costs, what are the social and organisational consequences?) to arrive at a decision about **acceptability** (on balance the best possible alternative for purchase)". Logan (1994) divides usability into two parts, behavioral and emotional. The latter means being desirable and serving needs beyond the traditional functional objectives.

Since smart products incorporate both physical and virtual dimensions, their level of usability depends on the design of both the hardware and the software. Similarly, their level of likeability depends on both of them. Designing a user interface is a multidisciplinary effort, and the target should be understood as a *whole-product user interface* (Vertelney and Booker 1990).

7.4   REPRESENTATIONS OF PRODUCT CONCEPTS

Iterative modelling and evaluation is a good tool for ensuring usability and likeability. The decision makers in a design process need information on which decisions can be based. The points of interest during the process shift from exploring the possibilities to managing risks. The various reasons for modelling and prototyping can be categorized into three broad classes (Säde 1996):

- idea generation,
- communication, and
- testing.

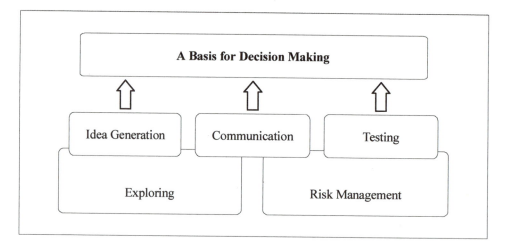

Figure 7.1 The reasons for modelling and prototyping.

The representations in the *idea generation* category include, for example, rough sketches for illustrating and evaluating design ideas. Sketching by drawing and 3D modelling is a basic tool for designers to generate, develop, and compare ideas. A designer uses visual thinking by seamlessly combining things he/she sees, imagines, and creates on paper. Once a sketch is on paper it leads to the next idea (Tovey 1989).

Communication is a natural function of any model. The builder of the model sends a piece of information mediated through the model to the receiver. A product development process involves experts from many different disciplines with different backgrounds, skills, and goals. They express their thoughts through different professional terminology. Concretions can support their communication by making abstract ideas explicit and understandable, which leads to a better product (Leonard-Barton 1991). Particularly, involving users in the process calls for concrete models, because they are not able to deal with abstract specifications. Companies also communicate through prototypes for purposes in marketing, training, and technology transfer, where prototypes serve as 'live' specification (Isensee and Rudd 1996; Kiljander 1997).

Testing is one of the most important reasons to prototype. The usability, as well as preferences concerning design solutions, can be tested and evaluated with real users. Nielsen (1993) sees that "the entire idea behind prototyping is to save on the time and cost to develop something that can be tested with real users". It is no exaggeration to say that the first and most important usability engineering technique is usability testing, whether it means small-scale user studies with rough prototypes at the concept creation stage, or formal usability testing with elaborated prototypes or semi-finished products in order to get the details right and to see if the goals have been reached. Prototypes can be tested with colleagues, other staff, different stakeholders, and naturally most importantly, with real users.

Models and prototypes can be of *high* or *low fidelity*. High fidelity models are finished and detailed, and they resemble the final product closely. Low fidelity models are limited in some way, they may be visually rough or represent only certain features of the product, etc. With a lo-fi model the aesthetic design of a concept cannot be evaluated as accurately as with a hi-fi model, but it is faster and cheaper to produce, which makes lo-fi models a reasonable choice in concept development. Virzi *et al.* (1996) have shown that usability problems can be effectively identified also with lo-fi prototypes, and under certain conditions they disagree with Nielsen (1990), who says that the type of usability defects found using lo-fi models differ from the ones found using hi-fi models. Schrage (1993) divides companies into prototype-driven and specification-driven ones. Prototype-driven companies tend to build lots of creative, lo-fi prototypes to drive the innovation process. He claims that they produce stronger products than the companies with rigid review structures, who build less, but more elaborated models. Sanders (1997) tells that at design consultancy Fitch, models appear nowadays 'earlier and earlier, and look uglier and uglier'.

7.5   SEVEN CLASSES OF REPRESENTATIONS

There are seven classes of representations that describe the emerging product, its user interface, and the interaction between the user and the product, which are described below (Säde 1997). Their benefits and limitations in modelling smart product concepts are listed.

Use scenarios are written stories of fictive persons using the product in certain situations. The character and background of a person are described with some detail, e.g. age, gender, profession, education, hobbies, likes, dislikes, and the level of skills with the use of technology. The story tells about the use of the product. The goal of creating several scenarios is to identify how different persons react to a given design in a given situation. The scenarios are usually supported by illustrations, and the persons and use contexts are preferably based on information gathered through studying real users.

+   The use of a product can be described from a detail level to a general level: the product, the interaction, the user, his or her feelings and values, and the social and physical environment. The description is free from constraints. Using scenarios, the designers are able, even forced, to consider the concept from a different perspective than their own.

-    As a model a scenario is separate from the product itself and abstract in this sense. It may be difficult to convince busy developers of the importance of scenario writing.

Drawings and storyboards include many kinds of two-dimensional visualisations from early sketches to photo-realistic renderings. They describe the emerging product or different parts of it, its user, its use, and its use environment. They may include supporting texts. Use sequences can be described through a series of pictures, i.e. storyboards.

+   Drawing is the basic tool for making design ideas tangible. Illustrations allow the evaluation of ideas by the designer him/herself or by others. Drawings make abstract thoughts or written descriptions understandable. Visualised ideas produce new, often improved ideas. Drawings support the generation, the comparison, and the recording of ideas, and facilitate communication.

-    Illustrations are two-dimensional, non-interactive representations of three-dimensional, interactive products.

User interface maps (UI maps) are diagrammatic representations of the structure of the user interface, and the possible paths of navigation. They may be abstract, with texts in boxes representing the states of the system (state transition diagrams), or include screen visualisations or thumbnail sketches of the appearance of the UI in different modes (UI maps).

+   UI maps illustrate the structure visually and can help to reduce complexity. They support project management and the design of consistent logic.

-    UI maps are abstract, understanding them requires expertise. They can also grow very complex themselves.

Physical 3D models are tangible three-dimensional mock-ups of a product. They can be built of e.g. cardboard, clay, foam, wood, plastic, or metal. Usually designers start by creating several rough mock-ups and end up with a finished appearance model.

+   Mock-ups are the only truly three-dimensional and tangible representations of the product and its UI's haptic part. They are understandable to everyone. They represent almost all the design related attributes: shape, colour, size, surface, weight, etc. Physical models facilitate analysis of the usability and style in relation to the human body and to the real use environment. They can be produced manually or from Computer Aided Design (CAD) files through rapid prototyping techniques, such as NC milling or stereolithography.

-    Physical models are static and non-interactive. A finished model requires a lot of human and financial resources, and is hard to modify.

Paper prototypes are low-fidelity prototypes of the UI for early usability tests. The user interface is drawn on paper and the changing UI items on separate sheets of paper or acetate. The user 'uses' the system by pressing the images of buttons etc., and one of the designers plays the part of the computer and manually changes the images of the UI items.

+   Paper prototypes are interactive and dynamic. The users can get a 'hands-on-experience' of the use situation. They allow usability testing before investing time or money. They are extremely fast and inexpensive to build. Even hard-to-implement or 'futuristic' features of the product's functionality can be effectively simulated with paper prototyping.

-    Paper prototypes are two-dimensional and low-fidelity, even if the illustrations were elaborated. Certain interface styles are hard to simulate, paper prototypes are best suited for a sequential dialogue. Feedback times are longer than in the final product.

Interactive prototypes are on-screen computer simulations of the user interface and the use of the product, which the user can manipulate via mouse, touch screen, or keyboard.

+   They are interactive, dynamic, and give a 'hands-on-experience'. Audio feedback enhances the face validity. A very real feeling of the final product can be achieved. Presents the relation of the fixed and changing UI items. Interactive prototyping is the only technique to evaluate feedback times, direct manipulation, text feed, etc. It allows usability testing, even at the keystroke level.

-    Interactive prototypes are always two-dimensional. There are problems in simulating products, which have controls on various sides or are too big to fit on computer screen, etc. The style of interaction is often unnatural. Many products require more than one finger for manipulation. Producing a high-fidelity simulation requires a lot of work. Mastering a scripting language is usually required. In the future it may be possible to build even complex interactivity without writing code, as the tools develop.

Computer modelling produces three-dimensional digital representations of the physical part of products. It provides information for manufacturing and realistic visualisations of the design.

+   The files are three-dimensional. Photo-realistic visualisations can be rendered. Physical models can be created through rapid prototyping techniques. The technique is well integrated with the Computer Aided Design and Manufacture (CAD/CAM). It is very precise and suitable for detailed design and assembly design. The tools are becoming more sophisticated, powerful and inexpensive all the time.

-    Even though the CAD files are 3D, the on-screen or printed output is two-dimensional, virtual, and not tangible. The models are not interactive, but software for building interactive 3D models will certainly be on the market in the near future. Computer modelling requires detailed input information, which may lead to too early commitment to the idea under development. This makes computer modelling not suitable for sketching and idea generation.

Using Table 7.1, Welker *et al.* (1997) identified which of the representation classes in (Säde 1997) described above, address the characteristics of users, of events of interaction, and of designed objects.

We see that all of the representation types naturally describe the emerging product. What is important, when interactive products are in question, is how to describe or study the use sequence. This can be supported with use scenarios, drawings and storyboards, UI maps, and paper or interactive on-screen prototypes. The most comprehensive focus can be achieved using scenario building and illustrations. Paper prototypes and interactive prototypes do not describe users, but are obviously among the most powerful tools for telling about users through studying them in tests.

In their article, Welker *et al.* (1997) also revised this list of representation types by separating storyboards from visualisations to a category of its own. This way describing events through storyboards is set apart from illustrations, which are static. However, placing storyboards and drawings into one group, even if they may consist of one or several images, can be justified, if we think of these two-dimensional illustrations from the point of view of the designer. When the designer sketches or draws, it is natural to draw several variations to describe the different states of the system, or to 'draw or write annotations' on an image to explain events in the use of the product.

Table 7.1 What the representations represent. Adapted from Welker *et al.* (1997).

Representations	Users	Events	Designed objects
1. Use scenarios	■	■	■
2. Drawings & storyboards	■	■	■
3. User interface maps		■	■
4. Physical 3D models			■
5. Paper prototypes		■	■
6. Interactive prototypes		■	■
7. Computer modelling			■

7.6 CONCLUSIONS

The iterative loop of idea generation, modelling, and evaluation serves as a tool for designing usable smart products. By studying the aesthetic design, physical ergonomics, and the cognitive usability through modelling a pleasant use experience can be designed. Modelling of these aspects should be done as early and as simultaneously as possible. Since the many dimensions of the use of an interactive product have an influence on each other, the designers should aim at modelling and evaluating the hardware and software parts of the product's user interface, the act of interaction, the user, and the context of use in a manner which makes this influence visible.

A challenge is to find a combination of techniques which flows from one to another in an integrated way, where the models provide as much benefit as possible. One prototype can be applied to several purposes, such as aesthetic design assessment, usability studies, and communication with stakeholders. The different techniques must always be applied to the situation at hand and they can be combined in various ways to find this synergy. Designers are aware of a set of techniques from which they pick one and use it their own way, which may vary from case to case. For example, an attempt to combine rough 3D mock-ups and paper prototypes is described in Säde *et al.* (1998). The designers ran initial usability tests with users in real environment using cardboard mock-ups with changeable UI items. The challenge in product development with real-world constraints is also to find a set of techniques which seems beneficial and justified in the eyes of management and those persons who are responsible for issues other than human factors. One way to handle this is to use well-focused, easy-to-implement, low-fidelity modelling techniques and to provide tangible results to inform the development process.

7.7 REFERENCES

Baber, C. and Stanton, N.A., 1992. Defining "problem spaces" in VCR use: the application of task analysis for error identification. In *Contemporary Ergonomics*, E.J. Lovesey (Ed.), Taylor & Francis, London.

Bauer, S.R. and Mead, P., 1995. After you open the box: making smart products more usable, useful, and desirable through interactive technology. In *Design Management Journal*, Fall 1995.

Brouwer-Janse, M.D., 1992. Interfaces to consumer products: how to camouflage the computer. In *Proceedings of CHI '92 Conference ACM SIGCHI*, N.Y.

Black, A. and Buur, J., 1996. Making solid user interfaces work. In *Information Design Journal*, **8**, 2.

Buur, J. and Windum, J., 1994. *MMI Design Man-Machine Interface*. Dansk Design Center.

Buurman, R., den, 1997. Designing smart products; a user-centred approach. In *Proceedings of IEA'97*, the 13th Triennial Congress of the International Ergonomics Association, Seppälä *et al.* (Eds.), Finnish Institute of Occupational Health, Helsinki 1997.

Caplan, S.H., 1994. Making usability a kodak product differentiator. In *Usability in Practice. How Companies Develop User-Friendly Products*, M.E. Wiklund (Ed.), AP Professional, Cambridge, MA.

Carr, D., Subrizi, A. and Faust, W., 1994. Reclaiming the soul of the synthetic object. *Innovation*, Summer 1994.

Feldman, L.P., 1995. Increasing the usability of high-tech products through design research. *Design Management Journal*, Fall 1995.

Isensee, S. and Rudd, J., 1996. *The Art of Rapid Prototyping*. International Thomson Computer Press, Cambridge, M.A.

ISO (1994) ISO DIS 9241-11 *Ergonomic Requirements for Office Work with Visual Display Terminals (VDTs):-Part 11: Guidance on Usability*. Draft international standard.

Johnson, G.I. and Vianen, E.P.G. van, 1993. Practical experiences with consumer products, users and prototyping. In *Contemporary Ergonomics*, E.J. Lovesey (Ed.), Taylor & Francis, London.

Keinonen, T., 1998. *One-Dimensional Usability. Influence of Usability on Consumers' Product Preference*. PhD Thesis (Doctor of Art), University of Art and Design Helsinki, Finland.

Keinonen, T., Nieminen, M., Riihiaho, S. and Säde, S., 1996. *Designing Usable Smart Products*, TKO-C81, Otaniemi, Helsinki University of Technology.

Ketola, P., 1997. Design aspects for mobile WWW browsers. In *Proceedings of ACHCI'97 Advanced Course on Human-Computer Interaction*, K.-J. Räihä, (Ed.), University of Tampere, Finland.

Kiljander, H., 1997. Interactive computer simulation prototypes in mobile phone user interface design. In *PB'97: Prototypen für Benutzungsschnittstellen. Notizen zu Interaktiven Systemen*. G. Szwillus (Ed.), Heft 19, Nov. 1997. GI-Fachgruppe 2.1.2. und des German Chapter of the ACM.

Leonard-Barton, D., 1991. Inanimate integrators: A bock of wood speaks. *Design Management Journal*, **2**, 3.

Logan, R.J., 1994. Behavioral and emotional usability: Thomson Consumer Electronics. In *Usability in Practice. How Companies Develop user-Friendly Products*. M.E. Wiklund (Ed.), AP Professional, Cambridge, MA.

McClelland, I., 1987. Consumer product design and the incorporation of ergonomics. In *Proceedings of INTERFACE 87. Human Implications to Product Design*. Symposium on Human Factors and Industrial Design in Consumer Products. Human Factors Society, Santa Monica.

Nielsen, J., 1993. *Usability Engineering.* Academic Press, Inc., Boston.

Nielsen, J., 1990. Paper versus computer implementations as mockup scenarios for heuristic evaluation. In *Proceedings   of IFIP INTERACT'90: Human-Computer Interaction.*

Norman, D.A., 1993. *Things That Make Us Smart.* Addison-Wesley. Reading, Massachusetts.

Säde, S., 1996. Modelling the design and the user interface of smart products. In *Designing Usable Smart Products, TKO-C81.* T. Keinonen, M. Nieminen, S. Riihiaho, and S. Säde (Eds), Otaniemi, Helsinki University of Technology.

Säde, S., 1997. Easy and pleasing - representing the design and the user interface of smart products. In *Proceedings of IEA'97 the 13th Triennial Congress of the International Ergonomics Association.* Seppälä *et al.* (Eds), Finnish Institute of Occupational Health, Helsinki, 1997.

Säde, S., Nieminen, M. and Riihiaho, S., 1998. Testing usability with 3D paper prototypes — Case Halton System. *Applied Ergonomics*, **29**, 1.

Sanders, E. B.-N., 1997. Vice President, Fitch, Inc. Columbus, Ohio, U.S.A. Interview on 21th March 1997.

Schrage, M., 1993. The Culture(s) of Prototyping. *Design Management Journal*, **4**, 1.

Shackel, B., 1991. Usability context, framework, design and evaluation. In *Human Factors for Informatics Usability*, Shackel, Richardson and Simon (Eds), Cambridge Univ. Press, Cambridge.

Scott, A., 1993. Product interface challenges — a blurring of distinctions. In *Design Challenges*, V. Popovic (Ed.), Queensland University of Technology.

Stifelman, L., Arons, B., Schmandt, C. and Hulteen, E., 1993. VoiceNotes: a speech interface for a hand-held voice notetaker. In *Proceedings of ACM INTERCHI'93 Conference on Human Factors in Computing Systems*, ACM 1993.

Tovey, M., 1989. Drawing and CAD in industrial design. *Design Studies*, **10**, 1.

Vertelney, L. and Booker, S., 1990. Designing the whole-product user interface. In *The Art Of Human-Computer Interface Design*, B. Laurel (Ed.), Addison-Wesley, Reading, MA.

Virzi, R.A., Sokolov, J.L. and Karis, D., 1996. Usability problem identification using both low- and high-fidelity prototypes. In *Proceedings of ACM CHI'96 Conference*, pp. 236-243.

Welker, K., Sanders, E.B.-N. and Couch, J.S., 1997. To understand the user/design scenarios. *Innovation*, **16**, 3.

CHAPTER EIGHT

Ergonomics and Safety in Consumer Product Design

DEVELOPMENT OF A TOOL FOR ENCOURAGING ERGONOMICS EVALUATION IN THE PRODUCT DEVELOPMENT PROCESS

BEVERLY NORRIS and JOHN R. WILSON

Product Safety and Testing Group, Institute for Occupational

Ergonomics, University of Nottingham,

University Park, Nottingham, NG7 2RD, UK

8.1 THE NEED FOR SAFER PRODUCTS

In the United Kingdom 4,000 people die every year and at least three million are injured seriously enough to seek hospital medical attention, all due to accidents in the home and at leisure (DTI 1995). Accidents in the home account for far more medically treated injuries (including those also treated by GPs) than injuries at work *and* on the road combined (DTI 1995). In 1989 the European Commission estimated that there were 30,000 deaths and 40 million injuries annually across the then twelve member states (Falke 1989) and the cost of home and leisure accidents was then estimated to be 60 billion ECU (Rogmans 1989).

Safety is an obvious criterion for a 'usable' product, all consumer products should be safe, efficient, reliable and durable (Kirk and Ridgeway 1970). An unsafe product is unusable in that it will certainly not be efficient or reliable and may not be durable either. Ergonomists suggest that consumers are becoming increasingly intolerant of poor design, and look for indicators of good design (Bullock 1994; Wilson and Whittington 1982). If we assume safe design to be a component of good design, this suggests that safety should now be a purchase criterion. With increasing competition and economic pressures any improvements in product safety could be a tangible factor in increasing market success which manufacturers would eagerly seize. However, even with product standards and legislation in place across Europe and elsewhere to ensure a minimum level of consumer safety, accident statistics indicate that there is still significant room for improvement in the safety of consumer products.

8.2 THE ROLE OF ERGONOMICS IN DESIGN SAFETY

It is proposed that there are a number of influences on consumer safety, as represented in Figure 8.1. The three immediate influences on consumer safety are the product design, the behaviour of the consumer (how the product is used) and the conditions in which it's used (where and when it is used and the product's condition). It is suggested that the latter two are at a 'micro' level of influences in that they vary from one instance of use to another and, beyond a certain level, are difficult to predict or control — an important aspect of consumer safety which distinguishes it from, say, occupational safety. The product design, its form, materials, construction and presentation, is the central influence on safety. The 'macro' influences on consumer safety can be thought of as those activities that are remote from the actual use of the product. They exert a direct influence on design, on the general level of safety in which the consumer market operates, and indirectly on the consumer. These macro influences are from government (in the general sense), product safety standards, industry, and, as a product of all these, safety information. Industry has an obvious role to play in the level of safety it decides to design and build into its products, the pressure it exerts on government, the level of safety it fosters generally in the market, and the influence it exerts on standards.

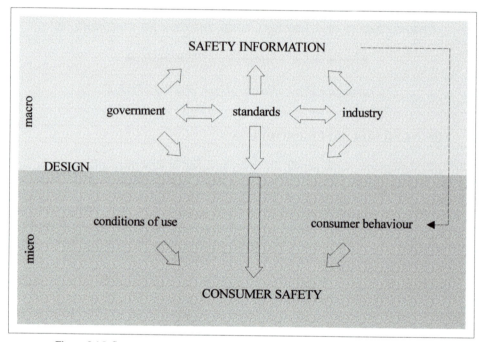

Figure 8.1 Influences on consumer safety.

At the root of all accidents is a product person interaction which has in some way failed, and either the product, person, circumstances or environment has precipitated this failure (Wilson 1983). Ergonomics has an important role to play in consumer safety by ensuring that this interaction is safe, and by improving product design, the central influence on consumer safety.

8.3   THE NEED FOR GUIDANCE

McClelland (1995) suggests that the fundamental principles underpinning ergonomics are based on the fact that the interaction between people and surroundings can be measured in a way that can be used to guide the specification and design of that interaction. It has been suggested that such guidance is either not available or not reaching product designers, especially in a form that can be applied to product safety (Wilson and Norris 1993). Ergonomics needs to be applied systematically and efficiently in the product development process if safety is to be ensured, and ergonomists should be providing tools to assist such application. This chapter describes the production of such a tool[1] — a publication aimed at persuading designers that ergonomics input to design safety can be a viable and affordable step in product development, and to both encourage and support ergonomics evaluation (Norris and Wilson 1998). The tool was conceived as a stand-alone document presenting designers with a framework for ergonomics evaluation, the target audience being designers and production teams working without any ergonomics support, and who would be unlikely to independently access information on ergonomics methods and techniques. The aims of the publication are to:

- promote safety as a design criterion,
- convey the relationship between ergonomics, safety and good design,
- make ergonomics and safety evaluation acceptable to designers and producers,
- encourage designers to see ergonomics as more than the use of basic ergonomics data,
- promote a structured approach to evaluating for safety, and
- provide advice on evaluation methods.

8.4 EVALUATING FOR SAFETY

Rationale

The concept of a step-by-step prescription of how to evaluate any design is attractive, both to ergonomists and designers. However, the conflict between generic and specific guidelines is a characteristic of ergonomics guidance, which makes it difficult to produce practical, stand-alone guidance across a large range of product groups. The requirement to be generalisable usually leads to a checklist format of 'ergonomics' factors to be considered (Ramsey 1985; Meyer 1994) but no real specific guidance on how to apply techniques. On the other hand, the requirement for specificity leads to a myriad of possible evaluation techniques requiring help in selection; bearing in mind that an important requirement of any design 'tool' was that it should convince designers of the feasibility of evaluation and not deter them with the prospect of investing time in selecting and becoming familiar with methods. There was thus a conflict between the need for completeness and rigour of any methodological guidance given, and for the accessibility and brevity of the publication.

A generic approach was taken, concentrating on the issues central to ergonomics and safety and providing a systematic framework for evaluation. This was based on the supposition that there are a number of basic factors to consider in *any* ergonomics evaluation and that these apply to *all* evaluations. Fundamental ergonomics issues are listed and illustrated, a framework for the evaluation process is presented and then the most commonly used methods and techniques summarised, to provide ideas on methods that designers could explore and adapt to their own needs.

---

[1] *Designing safety into products. 'Making ergonomics evaluation a part of the design process'* by Beverley Norris and John R Wilson, is now being distributed free of charge to industry by the UK Department of Trade and Industry.

Fundamental issues of ergonomics evaluation

Persuading designers to evaluate

Designers and producers have to be convinced that evaluating for safety is a viable and cost effective process which can be accommodated along with all other constraints on the product development process. Ergonomics has to be 'sold' before the choice of appropriate methods can even become an issue. The benefits of an established procedure, particularly in economic terms, need to be conveyed in that it can be used to improve all facets of the product design: usability and not just safety. The concept of 'evaluation of safety' is therefore broken down into tangible, recognisable units:

- ensuring a product is reasonably safe for its intended use and its intended users,
- extending such safety to include foreseeable misuse, including use by unintended users, especially the vulnerable such as children and the elderly,
- identifying those at risk,
- identifying the likelihood of injury,
- establishing the likely severity of any injury, and
- highlighting possible design improvements.

Evaluation as an integral part of product development

Designers have to see ergonomics evaluation as an integral part of the whole product development process, from initial concept development right through to monitoring product performance in the market. A model of the development process, together with its ergonomics inputs is shown in Figure 8.2.

The principles of ergonomics evaluation

With evaluation established as part of the development process, the fundamental principles of the relationship between ergonomics and safety need to be conveyed, to engender a user-focused approach to the design process. The features of the product person interaction which influence safety and which must be considered in any evaluation are described as (see Figure 8.3):

- Product, all features of the product should be assessed, including its structure, moving parts, controls, displays, casings, fixings, power sources, information and instructions, packaging and ancillary equipment.
- User, the physical and psychological characteristics of the likely user population (by 'user' we mean <u>anyone</u> who may come into contact with the product, either intentionally or unintentionally, and as a result of the primary or a secondary interaction with the product).
- Circumstances of use, how the product will be used, associated tasks and activities.
- Environment, the visual, auditory, physical and social conditions under which the product will be used.

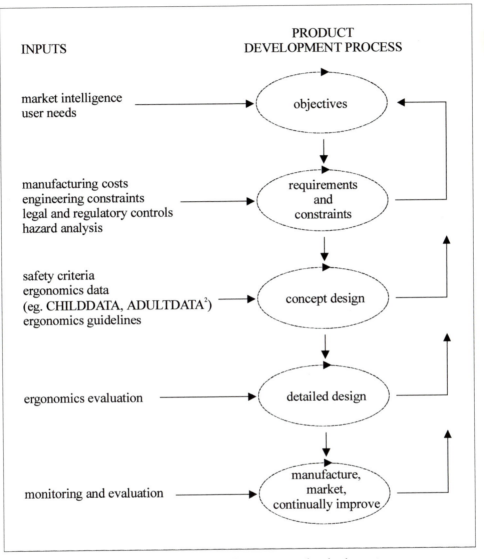

Figure 8.2 Ergonomics inputs to the consumer product development process.

---

[2] CHILDATA: The Handbook of Children's Measurements and Capabilities and ADULTDATA: The Handbook of Adult Anthropometric and Strength Measurements are designer resources to increase the use of ergonomics data in the design of consumer products. CHILDATA contains data on over 180 measurements on children, from birth to 18 years, including anthropometry, strength and product-related performance. ADULTDATA contains data on over 300 anthropometric and strength measurements on adults from around the world. Both are produced by the authors and distributed free of charge by the UK Department of Trade and Industry (see end for address).

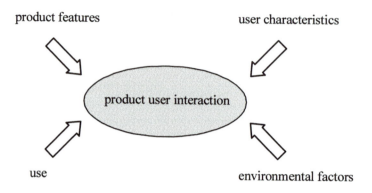

Figure 8.3 Factors influencing the product user interaction.

In order to stress the need to accommodate a range of users and their characteristics, some examples of the types of data that are available are given, as shown in Table 8.1.

Table 8.1 User characteristics to be considered in product design.

Characteristic	Examples
Personal characteristics	age, gender, literacy, education
Anthropometric characteristics	static body measurements *eg stature, weight, limb lengths, head size, hand size* dynamic measurements *eg reach, movement*
Strength	momentary or sustained forces that can be exerted or withstood *eg pushing strength, bite strength, manual handling*
Motor skills	psychomotor skills *eg hand-eye co-ordination, reaction times* gross motor skills *eg movement, balance, climbing stairs*
Skill, behaviour and performance	at tasks specific to a product
Sensory characteristics and tolerance	vision, hearing, taste, smell, touch, sensitivity to temperature, vibration
Experience and exposure	of a product, environment or task *eg years of driving, use of power tools*
Psychological characteristics	perceptual abilities *eg visual and auditory sensitivity* cognitive abilities *eg information processing related to labelling, attention, memory*
Personality	motivation, risk taking, perseverance, inquisitiveness
Socio-economic background	related to exposure and experience with products, tasks or environments *eg housing, income*
Cultural differences	physical, social, lingual
Disabilities	permanent and temporary conditions *eg pregnancy, fatigue, alcohol, drugs*

The importance of evaluation is stressed when any design change takes place, not just during new product conception. Evaluation is vital when designing a completely new product concept, a new product to fulfil an existing concept or redesigning an existing product, perhaps to incorporate a new function, feature, control/operating procedure, display, mechanism, structure or styling. The evaluation must consider all stages of the product's lifecycle: manufacture, transport, packaging, assembly, installation, use by all potential users, misuse by all potential users, storage, maintenance, cleaning, dismantling, disposal, re-use (second-hand life) and recycling.

An evaluation framework

As stated earlier a generic approach to evaluation hinges on providing a framework for the evaluation process, applicable across all products, hazards, and instances of use. A framework on which all evaluations can be based is proposed (as shown in Figure 8.4) consisting of four main stages:

Stage 1. Identification of all possible users

All possible users must be identified — intended, potential and inadvertent users — together with any differences between groups of users which may relate to their use of the product, such as those due to age (children and the elderly), disability, gender, literacy or cultural differences. Stage one is therefore a specification of users, using information from market research and accident statistics to produce lists of all users and of those potentially at risk.

Stage 2. Identification of all possible hazards

'Safety' needs to be transformed into tangible elements in order for it to be evaluated; these tangible elements are referred to as *hazards*. Different products can offer the same hazard, for instance those with the potential to induce a fall or entrapment, and so the same evaluation methods may be applicable to different products. A number of information sources and methods can be used to identify hazards:

- Standards and regulations, if they exist, will have identified some of the potential hazards and set minimum safety criteria.
- An analysis of accidents (both of accident rates and scenarios) will identify the common and less common incidents, who is being injured, where, when and how, and should account for the severity of any injuries incurred.
- Observing the product (or a similar product, prototype or similar task) being used by a range of potential users will give a demonstration of how a product may be used or misused.
- Potential hazard scenarios can be predicted by systematically building up all possible combinations of users, tasks, events and conditions.
- A design appraisal, by experts or users, may give an insight into potential hazards (this can be repeated as an evaluation method in stage 4 of the evaluation process).

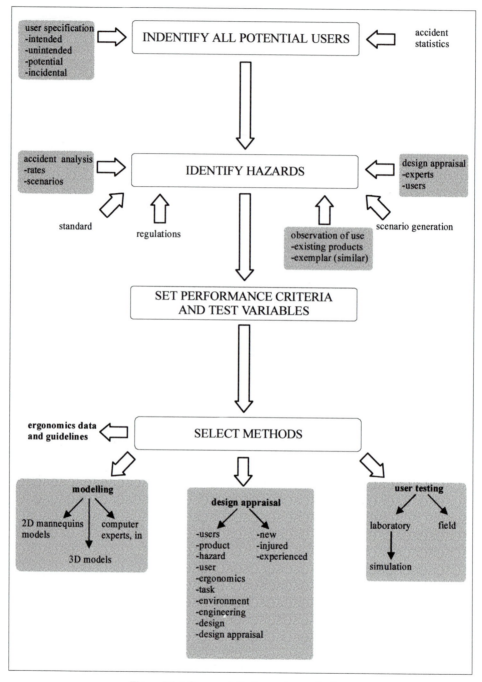

Figure 8.4 A framework for the product evaluation process.

Stage 3. Setting of performance criteria and test variables

Once hazards have been identified, performance criteria need to be set. These are design specifications which must be met in order that hazards are not realised. For instance, in the case of a safety gate a potential hazard may be the inadvertent operation of the catch by a child. The performance criterion would be that the catch cannot be opened by any child under five years. The test variables would be the amount of movement the child had made in the catch within a specified time limit. In some cases it may be difficult to 'design out' all hazards and a performance criterion may have to be set to include an acceptable level of risk, for instance a safety gate that could be opened by 10% of two-year-old children would be unacceptable but a 0.01% chance that an adult might cut their finger on the catch might be deemed an acceptable risk.

Stage 4. Selection of testing methods

Testing is a process in which observations of performance are assessed against the performance criteria. Methods of testing are numerous and can be concerned with technical as well as ergonomics aspects. Technical testing involves assessing performance of materials, electrical systems, mechanical systems and so on; ergonomics testing is concerned with any aspect of the design with which the user interacts.

Selection of methods

There is a plethora of ergonomics methods and techniques reported in the literature, and even more variants on the most popular methods. Once a method has been reported, it is often adapted for other applications. It has been found however, that a limited number of these methods are actually used regularly by experienced ergonomists (Stanton and Young 1997). It would appear unrealistic to expect designers and producers to navigate the ergonomics literature, to identify applicable methods and validate them for their needs. For these reasons, in the fourth stage of the proposed evaluation framework, ergonomics methods are classified into four broad groups or 'approaches': ergonomics data and guidelines, modelling, design appraisal and user trials. The use of these four groups of 'methods' was intended to help designers differentiate between the approaches to evaluation and to choose the most appropriate method. Within two of the groups a further range of 'techniques' is described, giving designers the ability to adapt the approach they have taken to their needs. Each group of methods and the techniques available within them is described, to stimulate new ideas and to stress the limitations of methods and any caution needed in their application.

Group 1. Ergonomics data and guidelines

...describing physical, motor and cognitive capabilities of users and general design rules derived from research on human performance.

The application of ergonomics to product design often begins and ends with reference to data and guidelines, and one of the aims of the guidance was to encourage designers to see the limitations in this approach and to encourage proper evaluation. The main limitations are:

- Inappropriate populations — differences in data due to culture, occupation, social class and secular trend must be considered.

- Generic versus specific data — the conditions in which data are collected have to be considered, particularly for performance or strength data.

Group 2. Modelling

...using some kind of representation of potential users at varying levels to assess a design, paper-based or computer-based.

Modelling has advantages over ergonomics data in that the evaluation can be specific to a particular product, safety tolerances can be calculated and interactions can be clearly visualised. Models are limited by the appropriateness of the anthropometric data they are based on. Also, and particularly with paper mannequins or 3D dummies, only a limited selection of the target population is represented. Computer modelling has obvious advantages in its compatibility with other computer design software, models that can be manipulated to some degree, re-used and adapted indefinitely, and evaluations are quick once models are established.

Group 3. Design appraisal

...using specialist knowledge to assess a design, with an insight that would not be provided by the design team alone.

Whilst design appraisal cannot be used to quantify hazards it can prove helpful in their identification and in stimulating the search for solutions. Specialist knowledge can be drawn from users (who can be experienced in the product, task, or environment; new or naive users or even those who may have been injured by a product) or experts such as ergonomists, safety professionals, engineers, other designers or people expert in the product (eg manufacturers, suppliers, installers, repairers, servicers), environment (eg school teachers, gardeners, residential care suppliers), task (eg parents, DIY enthusiasts) or the user groups (eg child care experts or carers).

Within this group of methods a variety of techniques are suggested such as: walk-throughs, focus groups, checklists, interviews and participative methods such as Design Decision Groups.

Group 4. User trials

...observing, measuring and assessing a sample of 'users' carrying out a task with a product.

User trials — any evaluation process where the user is represented in the test — are known to designers but are often considered too costly and time consuming or are carried out in only a limited form. The aim was to convince designers of the variety of techniques that could be used in user trials, such as simulation, in order to make them a feasible means of testing. Getting the most out of their investment was highlighted by stressing the importance of a number of factors:

- Subjects — ensuring they are representative, how to calculate appropriate numbers, recruitment, motivation, and instructions.
- Using simulation — of the user, product, task or environment - to combat ethical, practical, financial or time restrictions.
- Choosing the location of the test and the pros and cons of laboratory versus field testing.
- Selecting test variables.
- Measurement techniques — directly recording users' actions (observation, direct response, automatic event recording); self-records (questionnaires, diaries,

commentary, protocol analysis); recording opinion data (rating, ranking, interviews, checklists).
- Interpreting results (statistical significance, reliability, validity).
- Applying results.

8.5 APPLICATION OF THE GUIDANCE TO THE ASSESSMENT OF PRODUCT SAFETY

A substantial number of products are designed by independent designers or small consultancies working without any ergonomics support, so any guidance or tools they use have to be appealing and easily applicable. Prescriptive ergonomics advice for use by such designers is unfeasible, yet these are the very practitioners who most need ergonomics tools. Advice for designing for safety has concentrated on instilling basic ergonomics principles, conveying a role for evaluation throughout the product development process and providing a framework for evaluation that can be tailored to individual needs. Chapanis (1995), writing on the subject of using ergonomics methods in product design, has differentiated between ergonomics as a discipline and what ergonomists do. He defines ergonomics as 'a body of knowledge' as opposed to the practice of ergonomics, which is 'the application of that knowledge to the design of products'. He suggests that ergonomics is unusual as an academic discipline in that it has the aim not to just add to a body of knowledge but to apply it, but that only a minority of ergonomists are probably involved in actually applying ergonomics techniques to the process of designing or improving products. Early in the 1980s, Wilson and Whittington (1982) said that in the UK the discipline had failed to get ergonomics systematically applied in the design process, and that overt human factors input into the design of most new products was minimal, criticisms being the quality of research and the lack of adequate methodologies, techniques and data. Their recommendations to redress this were:

1. An integrated approach to design should be promoted and manufacturers convinced that ergonomists should be involved as early as possible in the design process.
2. An adequate interface between ergonomists and other professionals should be developed and communication improved.
3. Research findings should be generalised.
4. Research studies should be evaluated, for example for cost effectiveness, to improve methods.
5. Ergonomics data should be produced in a usable form.

It is hoped that the tool described in this chapter, and the approach it adopts, will work towards some of these recommendations — to improve the interface between design and ergonomics and convey the need for an integrated and systematic application of ergonomics methods to product safety.

8.6 ACKNOWLEDGEMENTS

The work described in this chapter was funded by the UK Department of Trade and Industry Consumer Safety Unit. Their contribution is gratefully recognised especially that of Paul Hale and Geoff Dessent.

DTI publications are available from:
R. Rose
CACP1
4.G.9
1 Victoria Street
London
SW1H 0ET, UK
Tel: 44(0) 171 215 0383

8.7   REFERENCES

Bullock, M.I., 1994. Ergonomics: a solution to product mis-match. Ergonomics: the fundamental design science. In *Proceedings of the 30th Annual Conference of the Ergonomics Society of Australia*, N. Adams, N. Coleman and M. Stevenson (Eds), 4-7 December 1994, Sydney, Australia, pp. 77-82.

Chapanis, A., 1995. Ergonomics in product development: a personal view. *Ergonomics*, **38**, 8, pp. 1625-1638.

DTI, 1995. *19th Annual Report of the Home Accident Surveillance System*. Department of Trade and Industry, London.

Falke, J., 1989. Elements of a horizontal product safety policy for the European Community. *Journal of Consumer Policy*, **12**, pp. 207-228.

Kirk, N.S. and Ridgeway, S., 1970. Ergonomics testing of consumer products 1. General considerations. *Applied Ergonomics*, **1**, pp. 295-300.

McClelland, I., 1995. Product assessment and user trials. In *Evaluation of Human Work: A Practical Ergonomics Methodology*, John R. Wilson and E. Nigel Corlett (Eds), Taylor & Francis, London.

Meyer, P., 1994. A product ergonomics evaluation checklist. Ergonomics: the fundamental design science. In *Proceedings of the 30th Annual Conference of the Ergonomics Society of Australia*, N. Adams, N. Coleman and M. Stevenson (Eds), 4-7 December 1994, Sydney, Australia, pp. 83-88.

Norris, B.J. and Wilson, J.R., 1998. *Designing Safety into Products — Making Ergonomics Evaluation a Part of the Design Process*. Department of Trade and Industry, London.

Ramsey, J.D., 1985. Ergonomic factors in task analysis for consumer product safety. *Journal of Occupational Accidents*, **7**, pp. 113-123.

Rogmans, W.H.J., 1989. Consumer interest in safety related standards for european consumer products. *Journal of Consumer Policy*, **12**, pp. 193-205.

Stanton, N. and Young, M., 1997. Is utility in the mind of the beholder? A study of ergonomics methods. *Applied Ergonomics*, **29**, 1.

Thomas, D.B., Dziambot, G. and Bohr-Bruckmayr, E., 1990. Ergonomics in product testing. *Ergonomics*, **33**, 4, pp. 453-458.

Wilson, J.R., 1983. Pressures and procedures for the design of safer consumer products. *Applied Ergonomics*, **14**, 2, pp. 109-116.

Wilson, J.R. and Norris, B.J., 1993. Knowledge transfer: scattered sources to sceptical clients. *Ergonomics,* **36**, 6, pp. 677-686.

Wilson, J.R. and Whittington, C., 1982. An overview of consumer ergonomics in the United Kingdom. *Applied Ergonomics*, **13**, 1, pp. 25-30.

CHAPTER NINE

Usability and Product Design: A Case Study

ROGER R. HALL

Department Safety Science, University of NSW,

Sydney 2052, Australia

9.1   INTRODUCTION

Various studies (Evans and Moore 1991; Thimbley 1991) have established that people have difficulty using consumer products and other modern technology, be they video cassette recorders, automatic teller machines or automatic ticket vending machines. However, despite an increased awareness in the literature about user centred design and designing for usability (Chapanis 1986; Eason 1992; Shackel 1986; Gould 1995) poorly designed products are still being produced.

Much has already been written about why designers get it wrong. Norman (1988) says designers go astray because the reward structure of the design community tends to put aesthetics first; Pheasant (1986) argues that putting aesthetics first is a fundamental fallacy in design. Many others (Thimbley 1991; Pheasant 1986; Norman 1988) argue that designers are not typical users and they have to please their clients who may not be the users. Also Pheasant (1986) and Ward (1994) argue that user involvement in the design process is perceived by designers as being unwarranted or too costly. Some authors however (Virzi 1990; Nielsen 1990) argue that significant design information can be collected from a small number of real users testing prototypes of alternate designs.

Stephen Ward, an Australian industrial designer, suggests that designers may feel that suitable methodology for user testing of alternative designs could be complicated and/or outside their expertise (Ward 1991). He goes on to state that "if simple methods can be readily employed by designers, these methods may be particularly useful early in the design process".

9.2   CASE STUDY

This case study examines two related Australian studies which examine the value of user testing of alternate designs of a domestic lighting controller. The studies examine usability issues and are based on the premise that involvement of real users early in the design process will provide useful design information at low cost and will result in a better and more usable design.

The first study (Ward 1991, 1994) used a simple cardboard mockup with limited fidelity and the second study (Kenney 1994) used a PC based touch screen prototype with higher fidelity. Kenney's study was conducted to confirm Ward's findings, but also to contrast the nature and extent of the design information found according to the level of fidelity involved. Before describing each of the studies and examining the value of the usability approach taken, the design and functionality of the original interface to the lighting controller needs to be detailed to provide the proper background.

9.3 DOMESTIC LIGHTING CONTROLLER

The domestic lighting controller examined was a microprocessor-based system which was scheduled for manufacture in 1991. It allowed any light in or around the house to be switched on or off from a master control panel. This panel would normally be located near the front door of a house so that if any lights were inadvertently left on, which could be seen from light emitting diodes (LED's) on the panel, they could be switched off by the occupants as they left the house. The lights could still be switched on and off as normal from a light switch in each of the rooms.

The controller had a capacity to control 32 lights and it could be programmed to turn nominated lights on and off at nominated times, to make the house look occupied at night. It could also be programmed to turn nominated lights on and off in response to infra-red sensors located in and around the house to automatically provide lighting of hallways etc or of the entry porch for returning occupants. It also had a security function which could be activated as the occupants left the house and, if an intruder was detected, all lights in the house would come on, hopefully to frighten the intruder away.

The controller had an internal clock which, together with the previously stored latitude and longitude, was used to calculate the times of sunrise and sunset so that the automatic lighting only operated during the hours of darkness.

The original design of the interface to the master control panel is shown in Figure 9.1. It has a 16-character, single line, liquid crystal display (LCD) and 20 pushbuttons which control up to 32 lights (or other electrically operated devices) and allow the system to be programmed as described above. The LCD display normally shows the time and date but also provides abbreviated programming instructions (because of the 16-character limit of the display).

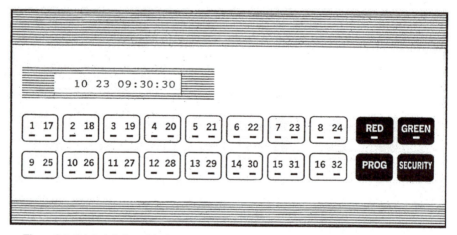

Figure 9.1 Original design of the interface to the domestic lighting controller. The little bars under the numbers on each pushbutton represent a red (left) and green (right) LED (from Ward 1991, 1994).

Sixteen of the 20 pushbuttons control the 32 individual lights and each of the pushbuttons has a red and a green LED embedded in it which provides feedback on whether the associated light is on or off. The associated light is indicated by the number label on the pushbutton immediately above the relevant LED. It can be seen from Figure

9.1 that the very first pushbutton (top left of the two rows) controlled lights No 1 (associated with the red LED, 'red') and No 17 (associated with the green LED, 'green'). The sixteenth pushbutton controlled lights No 16 'red' and 32 'green'.

Because each of the 16 light pushbuttons controlled 2 lights they had to be able to be put into one of two modes so that the 'red' and 'green' lights could be switched independently. This was achieved by the use of, not one, but two pushbuttons which toggled the 'red'/'green' mode of the light pushbuttons. These two pushbuttons were labelled RED and GREEN and respectively had a red and green LED under the label.

The two other pushbuttons, labelled PROG and SECURITY activated the Programming and Security functions of the lighting controller. When entering the PROGram function or mode the controller displayed short instructions for the user to follow, the exact details of which are not important here.

9.4 CARDBOARD MOCKUP INTERFACE

Original design

Ward (1991, 1994) utilised a simple cardboard mockup of the original interface which just had images of the pushbuttons but had a rectangular hole cut out of the cardboard exactly where the LCD display was situated (see Figure 9.1). By sliding cardboard inserts (each with a range of system messages written on it) behind the mockup, Ward could reveal the system response to the user through the LCD display cut-out. The user would touch the image of a push button on the cardboard face and Ward would 'display' the appropriate message.

Ward could also indicate which mode ('red' or 'green') the pushbuttons were in by sliding another card with patches of red and green colouring, behind the RED and GREEN pushbuttons which had very tiny holes cut out where the LEDs were situated. This enabled Ward to simulate which LED, red or green, was on and therefore which mode the lighting pushbuttons were in. Finally, Ward was able to simulate various states of household lighting by placing a clear acetate film, with small red and green bars representing active LEDs , over the interface to simulate which lights were on. If a light was off the respective LED was shown without colour.

It is one of the principles of ergonomics and usability that any product or system should be tested as a whole system and not just the interface. To this end Ward also allowed the users access to the set of instructions which was to be provided with the lighting controller. It is usual in usability testing to have users perform appropriate tasks to 'test' the whole system, and among other measures, to time how long users take to complete each task and to compile a list of errors made. In addition think aloud or verbal protocol techniques can be used to provide greater insight into what users' expected to happen and why they made particular 'errors'. Users were told that it was the product being tested and not them, so if they made an error it was because of a problem with the design.

Ward was not able to time how long users took to perform the set tasks and it could be argued that practising designers may also have the same difficulty if the testing is to be kept within certain resources. However Ward did compile a range of problems that users encountered and used a think aloud technique to provide insight into the associated design issues. He defined a problem as: 'any feature (or omission) in the product or instructions that was identified as contributing to:

- An error, any action that would have produced a system response other than that intended.

- Confusion, where the subject could not continue without assistance — even after referring to the instructions.
- Irritation or dislike, where a feature (or omission) in the product or instructions was identified as contributing to the subject commenting that a particular feature was difficult to use, or not the way they would have preferred — even if this did not actually result in an error'.

Ward tested 10 representative subjects over a period of a few days and found that there was sufficient fidelity in the cardboard mockup to reveal a total of 28 different problems with an average of 8.4 problems per subject (range 4 to 15) and a standard deviation of 3.26. Ward (1991) details the nature and extent of the 28 problems identified but several are listed below because their 'solution' is fairly obvious and could easily be incorporated into a revised design:

- They found the two button red/green toggle function confusing when selecting the lighting control mode.
- They found it confusing to enter numbers less than 10. The system required them to be in a two digit format, so 5 had to be entered as 05. This required the user to press the 10/26 pushbutton for the 0 digit and then the 5/21 pushbutton for the 5 digit.
- They found it confusing to refer to lights by numbers, although a suitable list could be prepared and kept with the controller.
- They found the American (month/day) format for entering and reading the date to be confusing.

Revised design

By examining and addressing the problems the users encountered with the original design, Ward was then able to design an alternate interface and quickly construct a revised mockup. This involved new and different 'hardware', 'software' and set of instructions. The revised interface is shown in Figure 9.2.

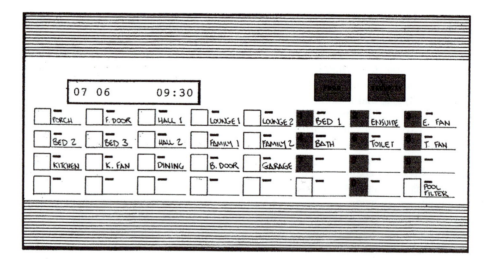

Figure 9.2 Revised design of the lighting controller interface. The little bar beside each pushbutton represents an LED to indicate the light is on (from Ward 1991, 1994).

The problems identified with the original design described above have been addressed as follows:

- A separate pushbutton for each of the 32 lights has been provided in the same space but with a slightly smaller area, this obviated the need for a mode switch.
- A numeric patch has been embedded to provide for unambiguous input of numbers
- A user definable label could be given to each pushbutton which could later be altered if the need arose.
- The usual day/month format for the date was used. Although this is not obvious in Figure 9.2, no one made a mistake setting or reading the date of 7 June.

Ward re-analysed the results from the ten subjects used with the original design and found that on average, six randomly chosen subjects from the ten would find over 80% of the problems. This is consistent with Virzi (1990) and with keeping the necessary resources down so that practising designers may be more inclined to use such testing.

Ward used only six new subjects to test the revised design and found a total of only 9 problems (including one new one) with the average number of problems per subject being 2.50 (range 1 to 4) with a standard deviation of 1.26. This represents a statistically significant 70% reduction in the average number of problems found after the design was revised.

While no specific costing of the time and resources involved in Ward's user testing was made, and while no actual benefit in the marketplace can be determined, it can easily be argued that a better and more usable design was a direct result of user testing of simple mockups within limited resources. It can also be argued that such user testing is, or certainly should be, within the capabilities of most professionally trained designers.

Touch screen interface

A touch screen interface can provide instant and very realistic visual and auditory feedback which Ward's cardboard mockups could not do. However touch screens cannot provide realistic tactile feedback but this can be implemented in a later iteration of the design if it is found to be necessary.

Kenney (1994) replicated Ward's study; however the cardboard mockup was replaced with a VDU based touch screen prototype which allowed greater interactivity between the user and the lighting controller system. The system messages shown in the LCD display were automatically generated by the system software in response to a user touching the image of a pushbutton on the screen. All LEDs were represented by screen images of appropriate size, colour and luminance. The time and date display was updated by the computer's system clock.

A life-size prototype of Ward's original and revised designs were created on a 14" NEC Multisync 2A monitor fitted with a Microtouch™ capacitive touch screen. The underlying functionality of the interface was implemented in the C++ object oriented programming language and is fully detailed in Kenney (1994).

Kenney also tested ten representative subjects on the original (and six on the revised) design and compiled a list of user problems (same definition as Wards). Kenney found a total of 54 problems with the original and revised designs — nearly twice as many as Ward found. This is not surprising, because the higher fidelity versions had an extra degree of interactivity afforded by the touch screen. This extra interactivity should not be confused with extra functionality as the touch screen simply provides better visual and auditory feedback.

Table 9.1 shows the average number (and standard deviation) of problems found per subject for both studies. It can be seen that both of Kenney's touch screen versions (original and revised designs) revealed more problems than Ward's low fidelity versions.

Table 9.1 Average number (and standard deviation) of problems experienced by users.

Study	Original		Revised	
Ward - cardboard	8.5	(3.3)	2.5	(1.3)
Kenney - touch screen	13.5	(2.9)	8.7	(4.9)

What is important in these results is that, had the original version been progressed to the higher fidelity stage without user testing, on average there would have been 13.5 problems per user compared with the 8.7 found in the higher fidelity, revised version. As it turns out, the higher fidelity revised design has approximately the same average number of problems as are found for the low fidelity, original version, 8.7 compared with 8.5. This indicates the value in user testing the low fidelity mockup and that the problems identified allow a significant improvement in the original design to be made quickly and at low cost.

Looking only at the number of problems is not always sufficient — some problems are worse than others and may severely limit the usability of the product. By defining a problem as major if more than 30% of users experience it, then out of a total of 54 identified problems found with the original design (both higher and low fidelity), 21 can be classified as major. Of that 21, the low fidelity version found 14 (67%) and the higher fidelity version found 19 (90%). However it must be pointed out that 6 of those found with the higher fidelity version could not have been found by the low fidelity version because of the restricted interactivity of the cardboard interface (i.e. no auditory feedback and only very basic visual feedback). Therefore the low fidelity version of the original design found 14 out of a possible 15 major problems, i.e. 93%. This indicates the efficacy of the low fidelity prototype in identifying most of the significant design problems.

9.5   DISCUSSION

Ward's cardboard mockup was made in half a day and took only some days to test ten representative subjects. The touch screen prototype was done as a final year, computer science project at the University of New South Wales (under the direction of the author). It took many hours of programming and testing but once developed it was easily altered to produce a second (and more if desired) design for evaluation.

By contrasting the results of both studies it has been shown, at least in the case of the domestic lighting controller described in this chapter, that most of the usability problems capable of being found by the cardboard mockup, were found. This demonstrates the efficacy of using low fidelity mockups with a small number of real subjects early in the design process to easily gather useful design information which can be used as a sound basis for redesign.

The same (and more) problems can be found using a PC touch screen based prototyping system. However, this involves greater resources (mainly time) and skill (programming) but may be seen to be worth the investment, especially in the modern PC based design office of practising designers. The problem still remains for the designer to be able to 'sell' the need for user testing to clients. A collection of case studies such as given in this book may go a long way to achieving this. Based on the results from the touch screen prototype, further improvements in the design of the lighting controller could be made and a fully working model (with real pushbuttons and LCD display) constructed and user tested. This would provide a basis to ascertain what problems users would have if the designs were taken through to manufacture. In this way they could be used to validate the usability approach described here as an effective and viable approach for designers to produce better and more usable designs.

9.6   REFERENCES

Chapanis, A., 1986. To err is human, to forgive, design. In *Proceedings of 25th Annual Conference of the American Society of Safety Engineers*, June 1986.

Eason, K.D., 1992. *The Development of a User-Centred Design Process: A Case Study in Multi-Disciplinary Research.* Inaugural lecture, 14 October, Loughborough University of Technology, UK.

Evans, M. and Moore, C., 1991. Technology: blessing or curse? *National Social Science Survey Report*, **2,** 5, Canberra, Research School of Social Sciences ANU.

Gould, J.D., 1995. How to design usable systems. In *Readings in Human Computer Interaction*, R.M. Baecker, J. Grudin, W.A.S. Buxton and S. Greenberg (Eds), (2nd ed), Morgan Kaufmann Publishers, San Francisco, pp. 93-121.

Kenney, T., 1994. *A Comparison between Different Interface Evaluation Methodologies: A Test.* Bachelor of Computer Science and Engineering thesis, University of NSW, Sydney 2052, Australia.

Nielsen, J., 1990. *Big Paybacks from "Discount" Usability Engineering.* IEEE Software, **7**, 3, pp. 107-108.

Norman, D.A., 1988. *The Psychology of Everyday Things.* Basic Books, N.Y.

Pheasant, S., 1986. *Bodyspace: Anthropometry, Ergonomics and Design.* Taylor & Francis, London.

Shackel, B., 1986. Ergonomics in design for usability. In *People & Computers: Designing for Usability*, M.D. Harrison and A.F. Monk (Eds), Cambridge University Press, Cambridge, pp. 44-64.

Thimbleby, H., 1991. Can anyone work the video? *New Scientist*, 23 February, pp. 40-43.

Virzi, R.A., 1990. Streamlining the design process: running fewer subjects. In *Proceedings of the Human Factors Society, 34th Annual Meeting*, pp. 291-294.

Ward, S., 1991. *Getting Feedback from Users Early in the Design Process: A Case Study.* A research project undertaken at the Department of Safety Science, University of NSW, Sydney 2052, Australia.

Ward, S., 1994. Getting feedback from users early in the design process: a case study. In *Proceedings of the 30th Annual Conference of the Ergonomics Society of Australia*, 4-7 December, Canberra, The Ergonomics Society of Australia, pp. 22-29.

CHAPTER TEN

User Trials as a Design Directive Strategy

WILLIAM GREEN and DAVID KLEIN

Faculty of Industrial Design Engineering, Delft University of Technology,

Jaffalaan 9 2628 BX Delft, The Netherlands

10.1 INTRODUCTION

This paper documents a project initiated by a major beer producer in the Netherlands. The company has a pro-active policy regarding occupational health and safety, and the problem of the manual handling of its 50-litre beer kegs has been a focus of attention for some years. Manual handling regulations increased the pressure to find a solution, and the original intention was to accomplish this by testing and improving a product idea that had already been developed into a second generation prototype. This product idea consisted of a hand truck equipped with a hook system to grab a keg and a tackle which was to be used by the lorry crews of the beer company. It was thought this product would be suitable to handle kegs in all kinds of ways during the delivery, but especially to be able to gently lower kegs into the cellars of the pubs.

Figures 10.1 and 10.2 (L) The original Keg Buggy and (R) the developed version lowering a keg.

However, a user trial with the prototype and the actual lorry crews, conducted at the delivery situations, revealed that the problems the product was supposed to solve were of quite a different nature than had always been presumed. As a result, the whole concept of the product idea had to be discarded.

By conducting a user trial of the distribution system, the real manual handling problems, their reasons for occurrence, alternative methods of handling kegs and the attitudes of the workers were made clear. This analysis of the problems made it possible to generate solutions ideas and choose those that were expected to actually solve handling problems in acceptable ways.

The manual handling problems were then connected to the appropriate solutions and the most effective development goal chosen to be researched further. Information from the user trials of the work with and without the aid of the Keg Buggy, and also from several extra tests, was integrated into this new development trajectory.

10.2 A SOLUTION FOR AN UNKNOWN PROBLEM

Earlier studies of the manual handling problems

As early as 1985 the management of the beer company had conducted research into ergonomic problems during the delivery of their products to pubs and especially of the delivery and retrieval of 50-litre beer kegs. Using electromyography, this research simply measured the extent and duration of bending during work. It showed that the work was quite tough, but not which methods of handling were causing the problems.

In 1992 the company called for a second study. This research consisted of analysing what handling methods were used during delivery, taking photographs of the situations, and analysing afterwards what kind of forces were minimally needed to do the work. This was done by modelling the situation and calculating the forces needed to maintain equilibrium. The derived values were then compared with Dutch and European laws and guidelines concerning work conditions.

This second analysis showed which handling methods were causing the problems, but lacked information about why workers used these methods. It was not known whether the workers had alternative methods, and if so, why they didn't use them. To be able to solve a problem in this context, one really needs to know, not only what the problem is, but also why it exists.

The development of the Keg Buggy

During 1995, a student of mechanical engineering spoke to the management of the beer company about an idea she had for making a hand truck especially suitable for handling beer kegs. She analysed what work currently had to be done with beer kegs: kegs had to be taken from wheeling containers, from pallets, transported over level floors to the pubs and often lowered into cellars. In agreement with the management, the so-called Keg Buggy was developed to do all these tasks. A first prototype was developed.

After a simple laboratory test, without using the lorry crews or the actual surroundings at pubs, the first prototype was found unsuitable for testing in the real situation, so a new prototype was developed. At the end of 1996, a second prototype of the Keg Buggy had finally been developed sufficiently to be considered suitable for testing with the actual lorry crews and during the actual delivery trips. The aim of the test was to seek out all the teething troubles of the design and find ideas for improvement.

At this point the project was taken over by a graduate student of the Faculty of Industrial Design Engineering at TU Delft.

10.3 USER TRIAL OF A WOULD-BE SOLUTION

Methodology of the user trial

A user trial of the proposed solution (the Keg Buggy) was conducted with five lorry crews, five drivers and four co-drivers, of different transport companies, and held in different regions across the Netherlands. In this way the lorry crews couldn't communicate and influence each others opinion before testing out the product themselves. It was also important to get a good view on the various load types, the different types of pubs, inter-regional differences and different negotiations with pubs by the representatives.

Each crew member was asked to try and find out the functions of the product themselves. When they didn't find out the basic functions after some time they were given clues. After acquiring a basic knowledge of the functions of the product this way, the crew was asked to load the Keg Buggy into the lorry. When, during the user trials at the pubs, it turned out that the workers had not discovered certain features of the product that were necessary or useful in using the product in an optimal fashion, more information was dispersed during the day.

Information of a general nature was gathered by giving the workers a questionnaire. They were asked how long they had been doing this kind of work, if they had ever suffered injuries, what kind, how long ago, what was the cause of these injuries and what the result of the injury was, such as absence, physiotherapy or relief from heavy duties.

The most important method of gathering information was to record on video the introduction, the work with and without the prototype, and the interviews with the workers. This was retained for subsequent analysis.

To be able to estimate the degree to which the Keg Buggy helped the workers to overcome manual handling problems, the workers were first asked to handle a few kegs in the normal way at each delivery situation, before trying out the prototype. In terms of methodology, it is interesting to note a dilemma between the interesting and sometimes important peripheral information picked up by the camera, and the need to reduce camera running time to keep the amount of footage workable. Video analysis can be very time consuming.

The researcher sometimes imitated the normal work methods of the lorry crews, as it is often difficult to get an impression of the physical difficulties the workers actually endure just from sight or from video analysis.

The video was sometimes used for analysis of the magnitude of the problems by determining the postures and calculating the necessary forces in equilibrium equations, as shown in the methods of the earlier research. The manual force used in some dynamic situations was also estimated by using the camera. For instance, when a worker pulls a keg off the top rack of a wheeling container and guides it down to the ground, he uses force to slow down the keg. By determining frame by frame (25 frames per second) at which moment the keg starts to move down and at which moment the keg reaches the ground, the drop time can be measured and therefore the minimally used catching force calculated. This method is quite crude and gives a rough but nonetheless useful indication.

The Keg Buggy as an all-round tool

It soon became obvious that the Keg Buggy was unrealistic as an all-round vehicle for handling kegs. It was thought that the buggy would be useful in taking kegs out of wheeling containers and from pallets, but this kind of work wasn't really a problem according to the workers. The workers would normally handle kegs by pulling, canting and rolling them by hand, with no, or hardly any, lifting of the kegs needed. These methods are very fast. Using the Keg Buggy is much slower and the buggy needs a lot of room to manoeuvre. In reality, there is often little room inside the lorry, or in the pub, to move about, causing the work with the buggy to be even more inefficient.

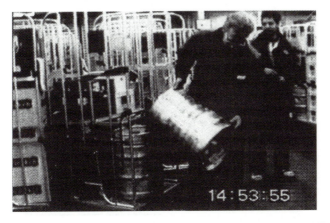

Figure 10.3 A worker pulling a keg from the top rack of a wheeled container and guiding it down.

For transporting kegs over level ground, the buggy was no competition for the small and lightweight hand trucks that are already in use. They are easily lifted from the lorry and need less room to move about. Besides, the hand trucks can carry two kegs at a time, while the buggy could only carry one.

Lowering kegs into cellars with the Keg Buggy

The most specific function of the buggy was its ability to lower kegs into cellars. It was equipped with a special hook system for kegs and a tackle. However, when comparing the work with the Keg Buggy to the work without it, the buggy was much slower. The workers had to position a keg, place the Keg Buggy against it, lower the hook system, attach the hook system, hoist up the keg, position buggy plus keg near the hatch, lower the keg all the way down, unhook the keg, hoist up the empty hook and move the buggy away for the next keg. This whole cycle took about a minute, even for experienced workers, while the normal methods take about 10 seconds. The interviews with the workers showed that even if the beer company were to give the workers more time to do the job, they themselves would rather work fast and have more time off in between to drink coffee, or to be home earlier.

It also turned out that lowering kegs into cellars wasn't such a big problem anyway. Normally one man would stand at the top of the cellar entrance to move the kegs through the hatch and the other man would stand below to move the kegs away. The detailed method used for lowering the kegs depended largely on the situation around the cellar entrances. In the best situations the workers would normally roll the keg in through a hatch and let the keg drop onto a 'drop cushion' — a slab of polyurethane foam with sufficient density to cushion the fall. The strength of the kegs ensures that they very rarely get damaged. This method of lowering the kegs doesn't require any lifting and is very efficient.

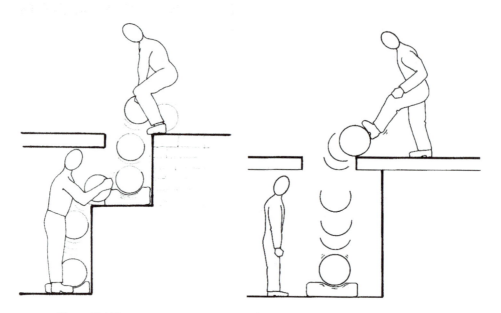

Figure 10.4 The current method for lowering kegs into a cellar: fast and efficient for most cellars.

Sometimes the particular situation in and around the hatch does make it necessary for the top man to lift a keg for part of a second, or for the bottom man to use too much force while controlling the keg and moving it to the ground. Often this has to do with combinations of hatches that are too small, obstacles such as ledges or pipelines under the hatch, etc. Only rarely were situations so awkward that the workers themselves thought they were a problem.

Conclusion

Besides its problems in fulfilling its main functions, the Keg Buggy had potentially dangerous stability problems and many minor development problems which still needed fixing in the design. However, those problems were irrelevant, as the most important conclusion of the user trial was that the whole concept of the Keg Buggy was completely unacceptable to the workers: the entire concept of using a tackle on wheels for lowering kegs was far too slow and the manual handling problems not, in the view of the workers, serious enough. To quote one worker: "I'd rather climb all the way up to the attic with a keg on my back than use that thing!" This demonstrates the power of work cultures, and reinforces the notion that instructions or rules only work if they are actually workable 'on the ground'. Otherwise workers will simply sidestep them. Similar attitudes were noted in a study of transport workers (Green and Barnett 1995). Such cultures are often only visible via trials of equipment or procedures.

The project then turned back to the foundation of the original objective of the Keg Buggy: solving manual handling problems in the distribution of beer to pubs.

10.4 CLARIFYING THE ACTUAL HANDLING PROBLEMS

To be able to develop solution ideas for some of the real manual handling problems, it was necessary to know more details about them. Happily, the eclectic manner of gathering information during the user trials with the Keg Buggy, even though considered perhaps overkill at the time, made it possible to generate the necessary information without actually conducting a second user trial.

The data were analysed and resulted in a complete list of the current manual handling methods that overstep the guidelines. It also summarised the alternative methods the workers have at their disposal, why they don't use those alternative methods and what, in detail, are the causes for the workers to overstep guidelines.

Some of the situations where the workers overstep the guidelines could simply be solved by handling instructions for the workers, when there were already acceptable alternative methods at hand. Others weren't so easy to solve. In particular, those handling problems that caused workers to overstep the guidelines and for which the workers currently see no good alternatives, were the basis for generating solution ideas.

10.5 SOLUTION IDEAS AND SIMPLE FEASIBILITY TESTS

Generating ideas

Several methods were used to seek solutions for the most pressing problems, as established during the user trials. Internet search engines, the library of the Delft University of Technology and the Dutch patent bureau were searched with keywords such as small transport vehicles, kegs, manual handling, hand trucks, tackling devices, etc. Over fifty companies specialising in small transport vehicles and lifting equipment were telephoned, faxed or visited. Two 'brainstorming' sessions were held with a diverse group. The ideas of the lorry crews were noted and evaluated. Mistakes and improvisation by the workers and problems encountered during the user trials were a valuable source for ideas.

The solution ideas that were gathered varied from solving the problem on a very fundamental scale, such as completely changing the way beer gets from factory to drinker, to changes to the kegs and small transport vehicles.

Figure 10.5 An example of an existing small transport vehicle: a hand truck which can mechanically climb stairs.

Idea selection

Although some good fundamental ideas were generated, none of them was expected to solve the problems of the keg distribution system in the near future, so the choice was made to focus on those ideas which would.

The solution ideas were connected to the problems they were expected to solve. By rating the severity of each specific type of manual handling problem, the estimated frequency of occurrence, the expected efficiency gain or loss and the total cost price of each solution and adding the scores, the most effective development goal was determined.

Using tests and user trials to evaluate solutions

Some of the information used in the evaluation was gathered by doing some simple tests with existing products which could have the capacity to solve manual handling problems, possibly after some adaptations. An electric stair climber was tested, a pneumatic tackle and an electric wheeling container puller. The wheeling container was first tested in a simple way at the distribution centre, adapted, and then tested again at the actual delivery situations. At all tests, the proceedings were again recorded on video for analysis.

The insight acquired during the first user trials into the Keg Buggy, slides, drop cushions, etc., was also integrated into the evaluation. This helped in estimating the feasibility of solutions. The understanding of the mentality of the workers played an important role.

Development goal

The choice was made to solve the rare but huge problem of full kegs that need to come out of cellars.* Full kegs need to come out of cellars when something is wrong with the keg or the beer, when pubs stop or change beer brand, when pubs find out they have ordered far too much beer, when pubs close for the season, etc. Due to the narrow and awkward cellar entrances, it is very difficult for the workers to get the kegs up. This was made very clear in the interviews during the user trials with the Keg Buggy.

10.6 DEVELOPING A TOOL TO IMPROVE THE MANUAL HANDLING CONDITIONS

Information from the user trials integrated into the design process

The Keg Buggy had sometimes also been tested for the function of getting kegs up from cellars, as well as being tested for lowering kegs. The lack of speed of the Keg Buggy was not such a big problem for this function, as the normal methods of getting full kegs out of cellars aren't very fast, but also because the workers view this type of work as a big problem themselves. In fact, some workers regarded the Keg Buggy as a reasonable solution for this work. However, there were stability problems with the Keg Buggy that would be hard to solve in order to make it an acceptably safe solution to the problem of getting full kegs from cellars.

---

* This is consistent with the approach taken in the transport study of Green, Barnett and Timmerman (1997), where a choice was made to focus on the cause of a relatively small number of incidents, but which, when they did occur, were very severe.

With the knowledge of the cellar situations encountered during the user trials, schematic lay-outs of cellars were drawn and the frequency of each situation estimated. Relevant details were:

- The size of the hatch; can the kegs go in horizontally or vertically.
- The type of support that can be achieved around the hatch: four sides, three, two opposite, two adjoining or even only one side free for support.
- The direction of the pull of the keg: straight down, partly forward (stairs or slide) or with a sideways component (spiral stairs).

Thanks to the data gathered on the videos, and the picture log book of the user trials, this analysis could be quickly conducted. The conclusion was that a product to get full kegs out of cellars should take into account situations with a forward component and should be stable even if only one side was free. The most difficult situation encountered was where a spiral staircase had no adjacent or adjoining sides free to use as support for a device. The pull of the keg is forward, sideways and changes as it drops. It is difficult to guaurantee stability in such a situation with any tackle device, and it occurs in about 20 % of the cellar situations, so can't be ignored.

When there is really only one side free for support, it means that any kind of solution that relies basically on a frame standing over the hatch would be very difficult to make suitable for a sufficient number of cellars.

Concept of an extra heavy buggy

In the end, two ideas were developed up to concept level: a heavy buggy and a portable tackle. The 'heavy buggy' uses its own weight and that of the operator as a counterbalance. It has an engine, a cable drum, a battery and two wheels for moving about.

Figure 10.6 Computer generated views of the 'heavy buggy'.

Thanks to two extendable supports, the moment of the operator and the weight of the product is increased compared to the moment of the keg. When the pull of the keg suddenly increases above a certain value, part of the crane will give way to gradually catch the fall and to minimise the effect.

Thanks to all these safety precautions it was calculated that this device will be safe when used properly. However, this concept would be a hassle to use and involve a lot of work to keep the product functioning properly: keeping the batteries full, checking the product regularly, etc. Maybe too many irritating problems for a product that will only rarely be of real use.

Concept of a portable tackle

The other concept was a simple hand tackle positioned on a portable lightweight crane. It has to be positioned in permanently placed attachments. The instalment of those attachments is the greatest problem of this concept: it needs to be done securely by professionals to ensure the necessary strength. However, when the instalment is done properly, using the portable tackle itself is very simple and safe. The tackle will take little room on the lorry and needs little maintenance.

Figure 10.7 Computer rendering of the portable tackle. It is light and easily transported to the site, or stored on the truck.

Extra research of cellar entrances

To estimate the complexity of installing the attachments, extra research was conducted at 25 cellars. The strength of the building construction around the hatches was investigated; if there was enough room for the crane to turn; whether the permanent attachment could be placed in a manner that wouldn't obstruct the normal use of the hatch. For instance, it would not be acceptable to place the attachment on the top step of a stairway, as it could easily cause people to stumble over it. If there is no other rim suitable, as in the case of a cellar with a stair directly below, the attachment will have to be recessed into the floor. That type of instalment will be relatively expensive.

The conclusions of this research were promising. It was estimated that in over 90% of the cellar situations placing and using the attachments will be relatively easy and therefore cheap.

Conclusion

Both concepts were evaluated and compared on the basis of the programme of requirements. The portable tackle was chosen for further development, particularly because of its safety and simplicity in use.

10.7 DISCUSSION

This paper shows how a development project can go off target when a problem isn't sufficiently well defined. The whole development of the Keg Buggy into two consecutive prototypes could have been avoided, or at least terminated in an earlier phase, if the manual handling problems had been user-trialed from the viewpoint of a problem solver. It would have been clear very quickly that the Keg Buggy was a solution for a non-existent problem.

The project also shows how conducting a user trial can be used for quickly scanning and evaluating possible solutions. By looking at existing products and testing them superficially, a better view can be acquired of their functionality. The designer can then estimate whether a solution idea or existing product can be developed so that it will solve the problem for which it is intended.

The method of conducting user trials can be used to test more than just products and people, but also to test circumstances, such as the situations at the different types of cellars during this project, and to determine their influence on the product.

A user trial can help to identify real manual handling problems, establish reasons for their existence, and help determine what types of solutions will assist.

The personal experience

Generally, the researcher conducting a user trial is expected to keep his/her distance from the actual testing, in order to maintain objectivity. However, in a circumstance such as the one described, no matter how useful the data captured on videotape, it is difficult to estimate how heavy or awkward a certain handling method is without personal experience. Therefore the researcher would sometimes imitate the workers in their handling methods. Of course it is still difficult to compare this personal experience of handling kegs a few times to that of workers who handle kegs day in and day out. However, the interviews showed that most injuries suffered by workers occurred when they were novices at the job, so in the end the insight acquired through personal experience was actually quite relevant.

In general, designers should not be afraid to improvise on the rules of conducting user trials, as long as they don't obstruct the objectivity of the research. For instance, during this project the researcher tried out handling methods shown by the workers, in which they are the experts, but refrained from testing the Keg Buggy in front of them and was economical in giving hints and guidance.

A user trial as a tool for generating ideas

The user trials helped in generating ideas (i) when seeing certain features in a product that might be combined with the features of some other product to make an optimal solution, (ii) when seeing the improvisations of the subjects in user trials, and (iii) most importantly, when viewing problems or things that go wrong. Most solution ideas generated in this project derived directly from the establishment of the real (manual handling) problems.

There is a also a danger in using a user trial for creativity purposes. The designer might get too focused on the tested object, try to solve all the little design flaws, and not be able to step back and look at the whole picture, or be inhibited from considering completely different alternatives. For instance, in this project the Keg Buggy was first regarded as the solution to the manual handling problems, but obviously failed the test. If it had not been so obvious, then effort might have been directed towards refinement of the design, and other alternative solutions might never have been considered. Many 'creativity' methods stress the point that idea generation should always start on a very abstract level, where all kinds of realistic objections should be forgotten. When ideas are generated on this abstract level, one tries to change some of these strange ideas into things that might work. Criticism is thus postponed.

A user trial, however, does the opposite. It is a tool to make the abstract concrete. Therefore, user trials should be used with caution in the early stages of a product development project, when one should be generating ideas instead of criticising them.

In this project, after the Keg Buggy had bee n discarded, several tests with products were conducted in a manner which did not obstruct creativity. By quickly scanning various existing products in a superficial way and near the end of the idea generation phase, the designer acquired information about solution principles of the test objects without becoming dedicated to a particular concept too quickly. These insights were very helpful later, when estimating the feasibility of the solution ideas to solve particular problems and in choosing the optimal development goal.

10.8 REFERENCES

Green, W.S. and Barnett, C.S., 1995. The ergonomics of heavy plant and equipment. *Report to the Government of South Australia Department for Industrial Affairs*.

Green, W.S., Barnett, C.S. and Timmerman, R., 1997. A case study in applied ergonomics and safety. In *Proceedings of the 33rd Annual Conference of the Ergonomics Society of Australia about Productivity, Ergnomics and Safety*.

CHAPTER ELEVEN

The User Interface Design of the Fizz and Spark GSM Telephones

BRUCE THOMAS

Philips Design, Gutheil-Schoder-Gasse 8-10,

PO Box 59, A-1102 Vienna, Austria

and

MARCO VAN LEEUWEN

Philips Design, Building HWD,

PO Box 218, 5600 MD Eindhoven, The Netherlands

ABSTRACT

The Fizz™ GSM telephone was introduced on the market in 1996, and has been praised in the independent press because of its 'unprecedented usability'. The Spark™ was introduced later in the same year and achieved similar acclaim, winning the *Media Total Mobile Phone of the Year* award primarily on the basis of the interface design. This paper describes the process by which this result was achieved, focusing on the activities of interaction design and human factors. The full integration of these disciplines was a key to the success, but a number of further issues are raised, including the need to adapt to changing requirements and the problems associated with the specification of innovative solutions.

Key words:

GSM, interaction design, interdisciplinary co-operation, product creation process.

11.1 INTRODUCTION

This paper presents a case study of the development of the Philips Fizz and Spark GSM telephones which were introduced on the market in 1996 (see Figures 11.1 and 11.2). In the development of these products, product management, hardware developers, software developers, interaction designers, human factors specialists and product management worked closely together with a user focus group.

Figures 11.1 and 11.2 (L) The Philips Fizz GSM telephone and (R) The Philips Spark GSM telephone.

The project is used to illustrate the process and the co-operation between different disciplines in the product creation process, with particular emphasis on the relationships between interaction design and human factors.

The techniques used during the development of these interfaces are an extension of those already described for the development of business phones (Thomas and de Vries 1995) and communications control terminals (Thomas and McClelland 1996).

11.2 THE DEVELOPMENT OF THE FIZZ

The development of the Fizz has been described in detail by van Leeuwen and Thomas (1997). A summary of the main points is presented here.

The Fizz GSM phone was the first hand-held GSM phone completely developed by Philips. Because it was a completely new product the whole development team needed to build up competence in the area of hand-held GSM phones. The design of the user interface took place over a two-year period. The low cost product was originally meant for consumers concerned with safety and convenience. Philips Design was involved from the start of the project.

User unfamiliarity with cellular telephones was identified as a potential problem which a successful interface would have to overcome. It was therefore decided to focus on the following objectives:

- It should be intuitive to make a simple phone call.
- The navigation structure should be simple and consistent.
- Telephone directory access should be simple.
- The interface should encourage exploration and use of more advanced functionality.

A constraint placed on the design was that in order to achieve low costs, both the display size and the number of keys should be kept as low as possible.

The following user tasks were assumed to have the highest priority for inexperienced users and should be most intuitive to access and operate:

- Gain access to use (insert SIM card, switch on, enter PIN code, etc.).
- Make and answer calls.
- Repeat last dialled number.
- Add and maintain items in a directory.
- Dial from directory.

Based on the first preliminary product descriptions from product management and the objectives and constraints mentioned above, two alternative interaction design concepts were developed with quick iterations. During this process paper, pen and flow charts were used to explain the interaction concepts to each other. Some initial simulations were made on the PC to get a 'feel' for and experience of the proposed interfaces. The simulation was used for internal discussions and resulted in further detailed alternatives.

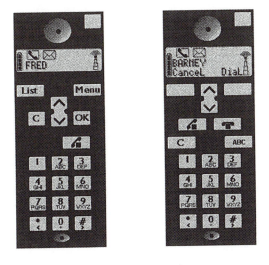

Figures 11.3 and 11.4 (L) Single line concept and (R) Soft key concept.

Two main interaction concepts were produced as a result of this highly intense and iterative process. The first was a single-line concept making use of a dedicated confirmation (OK) key for selection and activation, the second was a two-line concept making use of soft keys. The layout of the keys and the screen for the two basic concepts are illustrated in Figures 11.3 and 11.4.

Both concepts supported conventional dialling behaviour as well as cellular methods for dialling. (The user can 'go off hook' first before the number is dialled and can thus apply current experience.) A single-level menu was used to get a clear and simple menu navigation method for the intended target group. The menu was very long but ease of navigation was given higher priority than speed of access. Nevertheless, frequently used functions were placed at the top. The limited display space made it difficult to present useful text messages and options. To avoid text abbreviations, text was scrolled to the left when too long to be displayed.

The two concepts were simulated and tested in Eindhoven and Paris. There was no clear preference for either alternative, and the soft key alternative was developed further because of its overall flexibility. The test results indicated that in general users would learn how to operate the interface. After intense discussions with development and product management a number of changes were made. The OK and REDIAL key were added, the ABC key was removed and two separate telephone keys were combined (see Figure 11.5).

Figure 11.5 Improved concept after usability test.

Another major change was the introduction of an extra level in the menu structure. This was mainly because the number of functions requested by product management and demanded by the GSM specification became too many for a single level menu.

A further prototype was made and a further usability test with the new prototype was carried out in Eindhoven. The results showed that the design had been improved and successfully simplified. The test participants felt confident using the interface, and their comments were generally positive. The basic functionality was readily understood, as well as navigation in the menu. Simple dialling was achieved by all participants on the first attempt.

Nevertheless, improvements were needed. The route to add new entries to the directory was too complex. A better relation between soft key labels on the screen and the keys themselves was needed. Furthermore the cursor behaviour during text editing needed to be simplified and the message function was too low in the menu structure. There was a need for a general time-out, after which the system returns to the neutral state.

At this stage of the development process a more global approach was adopted. The management team decided that they would aim at a different target group, as a result of which they also changed the constraints within which the interface could be developed. The most obvious change was to drastically increase the display size and further to reduce the number of keys.

In order to develop a user interface within the new constraints and within the very limited time frame that was now available, it was decided to use the current user interface concept as much as possible. Thus the top row with fixed icons and two soft key labels were retained, while the main text area was expanded from one line to three lines with the introduction of improved feedback information. At the same time the key allocations were reviewed in order to reduce the number of keys. The OK and the telephone key were removed. It was recognised that this might make it problematic to make a call without the use of the telephone key. The product started to look like the final version (see Figure 11.6). Hereafter the complete interface details were specified on paper in the form of a Task Action Description and a simulation was made in parallel by the development team in Le Mans.

Figure 11.6 Third iteration.

The simulation was used for further testing in the Netherlands, Germany, and England. The results of this last test indicated that the test participants found the interface easy to use and were able quickly to learn how to operate it. A slight drop in performing the simple dialling task was observed due to the absence of the telephone key, but this was not as great as had been expected. However there were some usability issues that still required addressing, such as the use of symbols, particularly the repeat last call symbol, and the numbered structure of the directory. Users did not want a numbered names list and generally believed that newly entered names would be stored alphabetically. The task studies showed that finding the directory functions was still not clear enough. Further, the visual link between soft keys and their labels on screen needed to be improved and the text editor was still problematic. In the remaining time available it was not possible to implement all the recommendations from the test. Nevertheless, they were carried forward into the development of the Spark.

The main characteristic of the user interface is the relatively simple navigation structure using a limited number of function keys.

The product is now widely available on the market and has received positive feedback from the press about the ease of use. *What Cellphone* (1996) said: 'The menu system is very clear and easy to use and a nice operating touch comes when you make a call'. *Mobile Telecommunication* (1996) praised the excellent design and 'unprecedented usability' of the Fizz.

11.3 THE DEVELOPMENT OF THE SPARK

The design of the Spark user interface took place over a much shorter period than the Fizz. The decision to develop the product was taken in October 1995, that is before the Fizz was on the market, and the product was launched in August 1996. As with the Fizz, user interface design was involved from the very beginning of the project, and was considered by product management to be one of the central issues in the development.

The intended users of the Spark were to be mobile professional people. A familiarity with mobile phones was anticipated. On this basis, the main design objectives were to highlight quality, reliability and leading technology. The user interface was to build on experience gained from developing the Fizz, and should demonstrate innovation in a manner which supported the navigation structure already established.

The Spark introduced a large, full graphic screen of 64 x 100 pixels, together with a row of fixed icons above the pixel area. This enabled the display of up to five lines of text, and also enabled graphic enhancements such as icons, different screen fonts and graphical elements such as lines. In particular, the full graphic display enabled the use of animation which was used to support navigation through the interface structure.

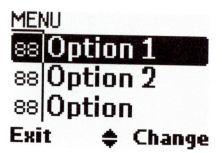

Figure 11.7 Spark interface concept.

The Spark user interface has nine function keys besides the 12 keys of a regular keypad. Two dedicated keys at the left of the product are used to control the speaker volume and one dedicated key at the right for switching the phone on and off. Five of the six function keys at the front are used for navigation, the other one is dedicated for the last redial number function. With the central up/down navigation keys the user can scroll upward and downward through a list of options. The soft keys are used in the same way as in the Fizz — they change their function depending on the location within the navigation structure. The 'c' key is used to make corrections and the redial key gives access to a list with the 10 last dialled numbers and 10 last incoming calls.

Most of the screens have a header at the top and soft key labels at the bottom of the screen. The fonts of the header, soft keys and content of the screen are different in order to separate the different types of information on the screen. This grouping of information is further supported by the use of graphical elements like lines and blocks.

Only one line on the display of the Fizz was used to display the option of the current level. The other lines on the Fizz were used to indicate the history by showing the selected option of each level within the menu. So if you are three levels deep in the menu two text lines will be used to indicate selections on previous levels. In the Spark we decided to show more options on the same level at the same time to give a better overview of options and only give history information about the previous level. For example, when the Menu option is chosen this will become the header and the options within the Menu group are shown in the form of a list. A maximum of three options can be displayed at the same time. The list starts to scroll smoothly when the cursor is moved toward an invisible option. The arrows between the soft key labels indicate that other options are available below and/or above the visible options.

This concept required a cursor and we decided to use an inverted black area for clarity and because it fitted in the graphical style developed by the screen graphics designer. The index number in front of the menu options is not only used to support quick access for expert users but also gives an indication to all types of users about the location within the navigation structure.

Animations were introduced both for aesthetic (pleasure, quality, fun, sparkling, surprise etc.) and functional reasons. We tried to avoid gimmicks and therefore used few amusing animations, although it was very tempting to implement all the animations originally proposed.

An animated welcome screen is shown directly after the phone is switched on. When the phone is switched off an 'inverted' animation is shown before the phone is switched off.

As stated before, the animations are used to explain the structure of the menu. Options from a group are presented in a vertical list. Within this list, the cursor moves smoothly up and down to the other items shown, or the list itself moves smoothly if the next invisible option is chosen. Soft scrolling is used to give a better indication of where you are going compared to the normally used 'jump' to the next option.

When the selected option is confirmed the next level of the navigation is entered. The display is cleared completely except for the selected option. This text element then moves in a very short time from its original position to its header position located top left of the screen. When it has arrived at its final position, the font is changed from lowercase to the uppercase font used for headers. At the same time the options for this menu level are revealed with a small animation. In one of our experiments we even scrolled the options from right to left but this was later thought to be overdone.

The sequence of the animation is reversed when a previous level is activated. The header moves back to its menu option position before the rest of the display is shown.

If the user wishes to go quickly to the next step of the interaction, all animations can be interrupted.

The idea of using animations created several types of response within the development team. Some reacted very enthusiastically while others immediately questioned the feasibility of the proposals and, together with that, questioned the usefulness of it. Fortunately, the key decision-maker believed in the concept and made it all possible. Nevertheless, many experiments and discussions were needed to find the appropriate animations. Experimenting with timing, smoothness, transition types and consistency of animations required a great deal of time and effort. It takes a lot of effort to find the right balance between expressiveness, smoothness and the duration of an animation.

Together with the screen graphic designer a large number of experiments were done to find the right balance between too much and too little information. The animation also needed to be consistent.

The menu structure and use of the soft keys built on the global interface and required improvements noted in the final Fizz test. The Task Action Description was again used as the main means to communicate the design to the development team.

A second version of the Spark introduced speech recognition for name dialling. This module was developed together with people from Philips Research Laboratories in Aachen. Ten names from the names list can be programmed to call by saying their name.

The user interface was simulated by the engineers in Le Mans, and a usability test was carried out in England. The test participants were drawn from a population of mobile executives, both with and without experience of using mobile phones. This test again showed that the concept was generally readily understood, and that most of the remaining issues from the last Fizz test had been satisfactorily resolved. An example is the now alphabetically sorted directory (names list).

The Spark user interface received even more critical acclaim than the Fizz, including the Media Totaal (1997) award for Mobile Phone of the Year. The reasons for selecting the Spark were:

'because of the nice, large display, the attractive menu control, the problem free choice between the small and large SIM card and the extremely effective and advanced speech recognition ('Voice Dial').'

The international specialist press (eg. Multi Media Practice Test 1997; Mobile Telecommunication 1997; Anderson 1997; le Marec 1997; Halliday 1997) issued several prominent reviews of the Spark, most of which focussed positively on the user interface and design.

The user interface of the Spark was recognized as one of the best, if not the best, GSM phone interface at the time. The size and weight factors are not the best but it is light enough to carry around. The product fits naturally in the hand and the waisting of the shape prevents accidental slip. One magazine mentioned that the keys need quite a firm push. Another magazine mentioned that the keys give a robust feeling, are well spaced have a definite click and are lighted in the dark. It was also mentioned that the ON/OFF key was difficult to operate. The display is described as big; bright and clear; able to display information; clear. One magazine mentioned the text font design specifically and judged it very positively. There is a good choice of ringing sounds and some are judged as very original. The loudspeaker was judged to be a tad quiet. All magazines are very positive about the menu system. They think it is logical, simple and clear, one of the best. The phone book (names list) is also judged positively because of its structure, looks and use. The animations are judged as slick, funny but also functional because it explains the hierarchical menu structure.

11.4 DISCUSSION

Not all the ideas generated during the development of the Spark could be implemented due to technical and time constraints. The time constraints and changing circumstances led to some misunderstandings during the implementation which could not be corrected in time.

Animations can be used to explain the navigation structure on products with small displays. They can also be used to increase the pleasure of use and ownership. Animations might give the impression of slowness because not all users will realize that they can interrupt the ongoing animation. This should be explained in the user manual and there should be an option in the menu to switch off the animation.

There are several problems with designing animations. Animation cannot be judged from a paper specification. At least a computer prototype must be made, but at present animations can only really be judged on the product display using the intended processor. Although first ideas could be prototyped with software like MacroMedia Director, the final processor and display characteristics greatly influence the final quality of the animation. Specifying an animation is difficult because of the timing aspects — you must feel and see it live before you can judge the timing aspects.

It is difficult to communicate innovative ideas because there is no existing reference for examples. This easily leads to misunderstanding and confusion. More time must be spent on the specification and the final implementation must be regularly checked. To get innovative ideas implemented and accepted, at least one decision-maker must believe in them and fully support them. This was one of the keys to the success of the Spark.

Finally, it should be noted that generally in the mobile telephone market there is a trend to produce ever smaller and lighter products. This provides a clear benefit of portability with comfort, but it means that the space available for interface elements is becoming more and more restricted. The Fizz achieved great comfort in the layout of the keypad due to the limited number of keys, a benefit which has enabled the same concept to be carried over to smaller products, including the Spark. However, as this trend continues, such concepts are likely to become unworkable and alternative strategies will be required (see Kaufman *et al.* 1996). This, then, will shape our co-operative work for the future.

11.5 ACKNOWLEDGEMENT

We would not have been able to achieve the results described above without the support and inspiration of our colleagues at Philips Design. We would like to thank, in particular, Pete Foley, Graham Hinde, Rolf den Otter, Bertrand Richez, Nicolas Ringot and Gus Rodriguez. Thanks are also due to the MMI Development team in Le Mans and the people from Philips Research Laboratories in Aachen.

11.6 REFERENCES

Anderson, S., 1997. *Spark One Up.* T3: Cool Gadgets and the hottest hi-tech hardware. April 1997.

Brigham, F.R., 1995. *Personal Communication.*

Halliday, R., 1997. Bright spark. *What Cellphone*, March, pp. 45-46.

Kaufman, L.S., Stewart, J., Thomas, B. and Deffner, G., 1996. Computers and telecommunications in the year 2000: multi-modal interfaces, miniaturisation and portability. In *Proceedings of the Human Factors and Ergonomics Society 40th Annual Meeting*, Human Factors and Ergonomics Society, Santa Monica, pp. 353-356.

Leeuwen, M. van and Thomas, B., 1997. The integration of interaction design and human factors in the product creation process: a case study. In *16th International Symposium on Human Factors in Telecommunications*, Oslo.

Marec, C. le, 1997. Spark de Philips: l'ergonomie grand public. *Telephone*, Fevrier, Zoom.

Multi Media Practice Test, 1997. Super light weighted. *Multi Media Practice Test,* pp. 22-23.

Mobile Telecommunication, 1996. Philips Fizzt to success. *Mobile Telecommunication,* **4,** p. 48.

Mobile Telecommunication, 1997. Philips Spark for big en small. *Mobile Telecommunication,* **1,** pp. 30-31.

Thomas, B. and McClelland, I., 1996. The development of a touch screen based communications terminal. *International Journal of Industrial Ergonomics,* **18,** pp. 1-13.

Thomas, B. and de Vries,. G., 1995. A user centred approach to the design of business telephones. In *15th International Symposium on Human Factors in Telecommunications.* Melbourne: Telecom Australia.

What Cellphone, 1996. Buyers guide. *What Cellphone,* November, pp. 67-85.

CHAPTER TWELVE

Usability Testing under Time-Pressure in Design Practice

ROEL KAHMANN and LILIAN HENZE

P5, Quality and Product Management Consultants,

Overtoom 283, 1054 HW Amsterdam, The Netherlands

12.1 INTRODUCTION

In many cases consumer products and products which need to be operated by consumers are developed from the point of view of technical quality and safety. The aspects of ease-of-use or usability tend to be neglected or left for the final phase of the development process.

P5, an agency which specializes in user research, has observed that contact with a new client often proceeds according to a characteristic pattern. The client only gets in touch after a product turns out to give problems in practice. They feel that something needs to be done, such as testing the product in a usability study to determine whether anything can be done. If the project is concluded and the product is improved, sometimes with high retooling costs, the issue is raised of avoiding the problem in a similar situation in future. In many cases, the decision is made to involve the user from the very start of product development.

Fortunately, product developers are becoming increasingly aware that an understanding of the interaction between user, product and environment can play an important role in the development process. One way of acquiring this understanding is to involve users in the various phases of product development. This is known as user-centred design.

Involvement of the user in product development can be effected by means of a usability study. When employed correctly, a usability study is a part of the product development process and is therefore subject to the preconditions of this process. This practice is characterized by restricted time and limited budgets.

This section considers the question of how the user can be involved in user-centred design. An approach to user research is described which is related to the constraints of product development, and the way in which it is applied at P5. Here, the emphasis is on process.

12.2 USER INVOLVEMENT IN THE DESIGN PROCESS

The production development cycle is becoming shorter and shorter (Schulze-Bahr 1993). This means it must be possible to set up and implement a user study within a short space of time. As a consequence, research methods need to be geared accordingly. Usability testing in the product development process demands a pragmatic approach rather than an academic line of questioning.

In order to carry out research effectively according to these preconditions, P5 has developed an approach during the past few years, which has a distinct place in the process of product development. It is assumed here that product developers farm out the usability study.

One argument for farming out the usability study is the objectivity aspect. A designer can evaluate his own work only with difficulty: 'nobody wants to kill his own baby'. The development of a product demands compromises with respect to the many factors that play a role. In particular, this process of compromise threatens in many cases to become opaque and rather casual. An objective evaluation can provide structure and thus has clearly measurable added value.

It is a fact that along with the usability aspects, many other factors play a role in the process of product development. Thus the product manager, engineers, designer and marketer will all try to mark this process in a way which is typical of their own angle of approach.

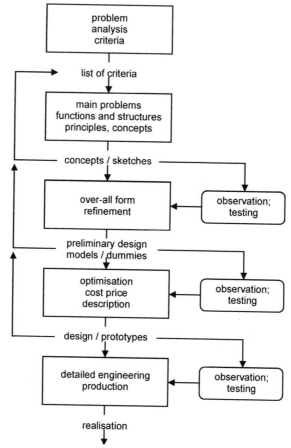

Figure 12.1 The process of user-centred design (from Den Buurman 1997).

The process of product development as a whole will not be considered in this context, since it is described exhaustively in general textbooks. A model has been adopted to serve as illustration (Figure 12.1) which is used at the University of Technology at Delft (Den Buurman 1997). We have limited this discussion to the usability study aspect. This part of the process can be described in the three main elements: briefing, research and workshop (Figure 12.2).

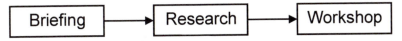

Figure 12.2 Three main elements of the usability testing process.

The briefing is the start of the procedure. In it the client describes the purpose of the usability study, including all preconditions such as information already available, the type of object, schedule, aims and expected results.

The research phase, itself composed of a number of steps, then follows. These steps roughly consist of a proposal with a provisional test design (tender); elaboration of the test design after being farmed out; organization of the test; recruitment of respondents; implementation of test; analysis of data; video editing in order to compile essential pictures and finally, the drafting of a report on the conclusions for the workshop.

One of the discussion points regarding testing is whether or not the designer may, or should, be present during the study. There are no objections in principle to his presence. A possible advantage here is that the designer may learn to see his product through the eyes of the user. This aim can to some extent be met by the videos which are made of all testing. This information has been compressed and can therefore be used more efficiently. One objection is that some designers have a tendency to fasten onto a few problems and to start working on them immediately. This makes it difficult for the usability tester to communicate the complete analysis.

At the end of this phase a workshop is held. This is attended by all the parties involved in the process. The choice of a workshop is quite deliberate. It is important that the information is communicated to the other disciplines. Any (preliminary) conclusions from this study which have already been reported can be explained and illustrated by such means as video. The workshop has proven itself an ideal tool for quick and efficient communication. It allows the time limits of the process to be met. The most important reason is the fact that we do not come up with solutions but provide arguments for finding solutions and making decisions. The aim of the workshop in this process is to employ as much information as possible from the usability study. This is a game that involves making compromises, since in many cases the maximum is not feasible. In conclusion, a (brief) report is written which carefully describes the decisions and the reasons behind them.

The entire procedure, from briefing to the workshop, is often realized within three to four weeks. This procedure can also be followed several times during the design process, a different 'subject' being studied each time. For example, the procedure can first be followed to generate ideas, then in order to test several concepts and, finally, a field test can be conducted on a prototype (see Figure 12.1).

12.3 THE P5 USESCAN® MODEL

P5's approach to this procedure is described in a model, the P5 Usescan®. The model is based on three elements: 'Object', 'Intervention' and 'Outcome' (Figure 12.3). These ideas are described briefly below.

The 'Object' is the subject of the usability study. Depending on the phase of the production development process, the qualities of the Object can vary considerably. These range from conceptual to material; that is, elements from the series, idea, concept, model, prototype and product. The model can also apply to non-material things, such as services, but this is a marginal point.

The 'Intervention' is the action which P5 or the usability tester carries out on the basis of the Object. The type of Object determines the aim of the Intervention. The aim, for instance, can be making requirements visible on the basis of a product manager's idea, or it can be to test concepts to determine whether they meet specific requirements. The aim also determines which methods are to be used and what types of data are to be obtained. In the case of generating requirements, use can be made of panel discussions and brainstorming techniques, while for researching a concept, the tendency would be to use a test configuration.

'Outcome' is the data which emerge from the Intervention in the form of information. How this information looks, and the form it takes, can vary substantially. In the foregoing example, in the case of requirements, the information would be a programme of requirements in the form of a list or report, while for the concept test, this can be an illustration of factors which determine use via a video tape.

The concept 'result' is deliberately not used here because a result can only be realized when the outcome is incorporated into the product development process and accepted. Communication with the persons involved in development, the design team, is very important.

A line is drawn in the model which defines when the object moves from the designer to the usability tester and where the data move from usability tester to designer (Figure 12.3). By doing this, the process can be repeated several times in succession. When the designer offers a concept, an adjustment will have to be made after the Intervention, and a more finely elaborated concept or model can then be tested again. This process can be repeated as often as necessary or desirable. Schedule and budget limitations, however, severely restrict such repetition in practice.

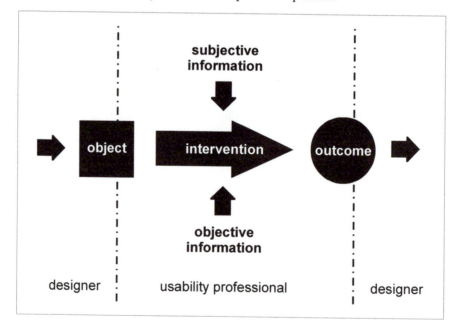

Figure 12.3 The Usescan® model.

The information which is used during the Intervention should be divided into objective and subjective information (Bonapace and Bandini Buti 1997). The objective information consists of facts and concerns information from product ergonomics. This is information which is primarily obtained through scientific research, such as anthropometry, bio-mechanics, cognitive ergonomics and standardization, and can be found in textbooks and databases. Such aspects concern dimensions, permissible forces, letter sizes and colour combinations. In short, aspects of a physical and cognitive nature.

Scientists provide other data for the benefit of the usability tester. This is information which concerns methods and techniques. Owing to the limited time available, the usability tester will employ 'standard' methods based on pragmatic principles. The academic usability tester can, and in our view should, compare various methods in a scientific study.

In addition to the hard facts available in the literature, there is also — and this is an important subject of this book — the input from the user. In usability testing qualitative methods are particularly used. These data are less absolute and can be considered subjective. Through keen observation and a structured analysis, however, these data too can be translated into more or less hard facts.

Criticism of user research focuses in the first instance on the limited size of the respondent group. In this framework, a parallel is usually drawn subjectively with the concept of representation from statistics. Representation is not about relating to a cross-section of society, but rather the contrary: the approach of P5 deliberately employs critical users. This means users whose view can be considered decisive to the question of whether or not a product or components can be used. The aim, after all, is to demonstrate and localize problems in order to eliminate them from the final product. Research has shown that the limited scope of the random sample is effective in uncovering the majority of problems (Virzi 1990 and 1992; Dumas 1997). The remaining problems can only be uncovered through great effort, so further work no longer justifies the expense (the familiar 80/20 rule).

In brief, the model gives insight into the way usability testing functions in product development, and makes it possible for the process of briefing to be followed to the workshop phase. Linked to this, P5 distinguishes three types of Interventions.

12.4 INTERVENTION: THE THREE PHASES OF USABILITY TESTING

In terms of the design process, there are three different types of Intervention possible: analysis, testing and verification. This situation applies to new and existing products alike. We call these the three phases of usability testing. It should also be pointed out that when the product is complex, more Interventions can take place, such as when a product is tested several times.

Analysis

It is important to determine which aspects play a role in (future) use. The purpose of the analysis phase is to identify the use parameters. Three elements are concerned here: the product, the user and the context in which the product is used. The parameters are translated into a programme of requirements used for the concept development.

The point of departure associated with the right application of the principle of user-centred design lies in the designation and understanding of the target group, and in the careful deliberation about the group of end users. Is the product intended for young persons, older persons, professionals, or for everyone? The user-centred approach is characterized by the criteria 'understanding the users' and 'involving users in product development' (Pacenti 1997). There is nothing wrong with excluding certain groups of users. What is important is to remain conscious of this aspect and be aware of the consequences beforehand.

Certain users can also be covered deliberately. When, for example, usability testing is carried out for general consumer products, such as washing machines, microwave ovens or telephones, it is certainly worth involving, along with the 'average' user, older people as critical respondents (Butters and Etchell 1997).

A product has a kind of presence; an appeal. This makes emotional aspects relevant and they can vary completely from one age group to another (Jordan 1997 and 1998). The effects of environmental factors should also be borne in mind. It should be realized that a product is used in a certain context. The signal of a telephone in a sitting room is expected to meet different requirements from one in a noisy factory.

Cultural effects are also important. A European study of mobility indicates that elderly Fins consider it important that the countryside is within reach, while elderly Italians attach the greatest value on being able to reach family and friends (Mollenkopf 1997). These are matters that, along with functional aspects, play an important role in the implementation of usability testing. It should therefore extend further than the mere testing of models in a usability lab.

In this phase it is important to determine a so-called 'interaction profile'. This is a description of all relevant interactions with the product. A user profile is then compiled in which all relevant information of potential users is recorded. From this information, the critical users, the critical user aspects and the critical user circumstances are determined.

For existing products a number of methods are available for obtaining this information. The choice of method is in the first instance determined by the research question. In practice, however, many methods have proved to be time-consuming and it is often time which is in short supply. For example, the drafting and implementation of a questionnaire may be too labour-intensive and therefore not feasible.

During the analysis phase we usually conduct desk-research in combination with a panel discussion (focus group) or expert interviews. Desk-research is aimed at data from the manuals, ergonomic aspects of the product and standards and guidelines associated with the product as well as cognitive ergonomics. Information is also compiled which is required for identifying with the topic of the study.

Focus groups and expert interviews have proved to be very efficient methods of obtaining large amounts of information within a short time. When the research concerns a product which is not yet available, then use is made of brainstorming techniques to generate new ideas, in addition to the aforementioned methods. The information in the usability testing is laid down, along with other relevant data, in a programme of requirements.

Testing

On the basis of the data which emerge from the analysis phase, the designer or design team go on to develop concepts. During this process they can check the programme of requirements to determine whether the concepts comply with the requirements which have been set.

During this second phase, one or more concepts are tested according to such criteria as effectiveness, efficiency and satisfaction. The users comment on these points. In most cases a protocol is drafted which lays down the entire test procedure. According to the product, a number of techniques are used for implementing this study. The best known is the laboratory test, in which use is made of a lab with a one-way screen. The respondent plods away silently on one side of the mirror with the researchers and designers sitting on the other side.

The controlled circumstances of a laboratory test have many advantages but also involve the disadvantage that the actual circumstances of use are partly, and often entirely eliminated. When circumstances play an important role, a field test is used. A field test gives a more faithful picture but confronts the user with the substantial disadvantage of being unable to control all circumstances. Experience has shown that the tester is relatively dependent on the co-operation of others. This situation requires considerable organizational efforts.

During laboratory research P5 seldom uses a one-way screen and conducts research much more intensively, that is, the researcher is present in the room with the respondent. In addition to this form of individual testing, panel sessions and testing in pairs (co-discovery) are also possible. This research is qualitative and employs a relatively small number of respondents.

The product being tested can have various forms. These include paper-based mock-ups, software-based mock-ups (testing with a touch-screen for instance), three-dimensional mock-ups, prototypes, or complete products. A separate form of this testing is comparative testing. Here, several models and brands of the same product are compared to one another. For such testing, larger panels are generally needed.

Verification

Verification is the concluding phase. The term 'evaluation' is often used here. This phase can serve a variety of objectives. A definitive concept, for example, can be tested before being elaborated further. Another possibility is to test the reliability of the final version, such as when the condition is set that 95% of the users should be able to use the product without problems the first time. Verification is quantitative and is implemented with large groups of users.

In most cases observations or structured interviews or surveys are used. Through the use of large numbers of respondents elements can be tested in this phase such as preferences, pricing and the like. It will be clear that the field of the market researcher is sometimes involved here. The techniques are similar.

12.5 HOW TO FIND USERS

When the decision is made to involve users in the design process, these persons must then actually be found. There are various ways of approaching respondents. The most important are the following:

- Employ an agency to find respondents. Such agencies are able to produce the required number of respondents within a short time. The disadvantage of these agencies is that they register their respondents primarily according to social or demographic features and not according to the functional characteristics that relate to the product. And it is precisely a sound selection made according to functional features that can be important for an effective test.
- Place a notice in the newspaper, national or regional according to the desired reach. The ad should be along the lines of 'Research agency needs users of product X between the ages of 55 and 70' or 'Have you recently purchased a brand Y camcorder? We're looking for...' This method is particularly suitable when highly specific requirements are set on respondents. The disadvantage is that the response is unpredictable and little is known beforehand about the respondents.

- When users are required from more specific groups, it is possible to recruit them from umbrella organizations, interest groups, associations or vocational groups. For example, in the case of professional products or products for people with a certain handicap. Another possibility is to carry out the test during trade fairs, should they occur at the right time.
- Owing to the speed at which assignments are set up and carried out, P5 frequently uses its own panel database. This means, however, that the organization and management of the panel will take much time. Points of attention here are updating information, the protection of privacy, call frequencies, the demand made on panel members' time and the composition of the panel.

12.6 CONCLUDING REMARKS

User-centred design is a process of usability testing by which designer and producer obtain an objective assessment of factors that play a role in the use of the product.

The short cycle of product development has resulted in product testing being dominated by time limits. This situation requires methods which combine pragmatic and scientific principles.

Much has been written in the literature about the various methods that can be used for usability testing. Very little has been written, however, on the role of the usability tester in the greater totality of the production development process. We hope that the description of the P5 Usescan has made it clear that, in practice, usability testing may readily serve as an integral part of product development. It is a method which is aimed at limited budgets and limited time.

In addition, we hope the point is now clear that usability professionals and scientific researchers need each other in order to develop usability testing further. Professionals make use of data from scientific research. Scientists can develop methods further on the basis of questions drawn from practice.

12.7 REFERENCES

Bonapace, L. and Bandini-Buti, L., 1997. The pleasant object: The Sensorial Quality Assessment Method. *Ergonomia*, **5**, 10, pp. 11-14.

Butters, L.M. and Dixon, R.T., 1998. Ergonomics in consumer product evaluation: an evolving process. *Applied Ergonomics*, **29**, 1, pp. 55-58.

Butters, L. and Etchell, L., 1997. Design for all: evaluation for all. In *Proceedings of 13th Triennial Congress of the IEA*, **2**, Tampere, Finland, pp. 202-204.

Buurman, R. den, 1997. User-centred design of smart products. *Ergonomics*, **40**, 10, pp. 1159-1169.

Dumas, J., 1997. How many participants in a usability test are enough? *Common Ground*, **7**, 4, October, pp. 2-6.

Jordan, P.W., 1997. The four pleasures — taking human factors beyond usability. In *Proceedings of 13th Triennial Congress of the IEA*, **2**, Tampere, Finland, pp. 364-366.

Jordan, P.W., 1998. Human factors for pleasure in product use. *Applied Ergonomics*, **29**, 1, pp. 25-33.

Mollenkopf, H., 1997. The everyday activities and mobility of elderly people: needs and hindrances. In *Proceedings of 13th Triennial Congress of the IEA*, **5**, Tampere, Finland, pp. 600-602.

Pacenti, E., 1997. *Paper Presented at Presence Forum*. Royal College Of Art, 6 November, London.

Schulze-Bahr, W., 1993. Designer must engage in dialogue with consumer. *EDC-News*,
  1, 4, pp. 4-7.
Virzi, R., 1990. Streamlining the design process: running fewer subjects. In *Proceedings
  of the Human Factors Society 34th Annual Meeting*, pp. 291-294.
Virzi, R., 1992. Refining the test phase of usability evaluation: how many subjects is
  enough? *Human Factors*, **34**, pp. 457-468.

Case Study

USABILITY TESTING IN THE DEVELOPMENT OF A COFFEE DISPENSER THE CAFITESSE 400

The Cafitesse 400 is a compact, professional coffee machine especially developed by Douwe Egberts Coffee Systems for large-scale consumption and the catering industry. The machine is capable of producing several types of coffee at a rapid rate (Figure 12.4). After its introduction, Douwe Egberts also needed to use this machine in a self-service environment as a dispenser. Because the user associated with this situation is of an entirely different type, P5 was asked to analyze the consequences of this plan.

It was decided to carry out a usability evaluation to determine possible problems and potential users. The client's primary precondition was that the dispenser should be modified as little as possible because of market introductions already planned for it.

The survey was implemented in three phases and took place in consultation with the client and the designer of the Cafitesse 400, the Neõnis Industrial Design agency at Zaandam. The project illustrates the approach and methods of P5 which are based on the Usescan method. The three phases of the project were (1) analysis, (2) concept test and (3) verification.

Analysis

The analysis was aimed at determining which aspects play a role in use and which users are decisive. During the briefing the client made it clear that the product concerned was to be sold world-wide.

The method used to compile information was a field study conducted at locations where this dispenser is already used in a self-service context. Observations were made on site. In addition, short interviews were held with people who had just drunk coffee. The personnel at each location were also interviewed. The locations, five in number, were in the Netherlands, Denmark and the United Kingdom. The selection of the locations was in this case based on the presence of the coffee dispenser.

This analysis made it clear that the placement of the cup and the location of the right push button are elements which are decisive in the use of the dispensers. The dispenser should indicate a clear relationship between the control panel and the place where the cup is to be placed. Findings were discussed at a workshop with a production line manager, engineers and the designer.

Concept test

The conclusions of the workshop were used by the designer to develop three different concepts. The concepts were evaluated by a user panel in individual sessions. Two factors in particular were taken into account. First, it was determined whether respondents indeed recognized the buttons on the control panel as push buttons. The analysis showed that this aspect may well be a problem. It was revealing that in practice stickers were used giving such directions as 'Push here once'. This practice is usually an indication that something is wrong.

The respondents were shown a number of models with push buttons and direction plates, including several new concepts for the push buttons. The respondents were then asked to indicate whether they thought the model shown was a direction plate or push button type.

Secondly, the location of the cup was considered. The Cafitesse 400 has three flow openings. This means that the cup can be placed in any of three different places. Several configurations were used to arrive at the cues people use for determining the cup position. A general convention emerged where the cup is always thought to belong in the centre. This is the case with a great many beverage dispensers. A number of models of the rack (tray) on which the cup is placed, and of the cover with the outflow openings (the brew basket) were tested. A few complete machines were also set up in different configurations. Respondents were requested to take a certain type of coffee. The data from the various test components were again analyzed in a workshop. On the basis of these results the client and designer made their definitive modifications.

Verification

The modifications were built into a prototype. That is, a new model was made of the push buttons, the brew basket and the tray. These parts could easily be fitted on an existing coffee dispenser. This prototype was used for the third phase, verification (Figure 12.5). The client required that with the new solution, the use of the dispenser in a self-service situation would be problem-free in 95% of the use. To test this parameter, the model was placed at two locations in a self-service environment and a quantitative test was conducted which scored 'good' or 'faulty'. These tests achieved a favourable result.

Conclusion

The study clearly demonstrated that the application of the original model of this machine in an inexperienced user environment is not feasible. Involvement of the user made it possible to obtain the information required for optimizing the product.

The importance of the user research does not lie in the ability to create solutions for improving the product, but to facilitate the formulation of required solutions. The development of creative solutions was left to the designer. The researcher should be considered a sparringpartner of the designer and not, as many designers initially fear, a bogeyman who lays down the law.

Figure 12.4 The original version of the Cafitesse 400.

Figure 12.5 The prototype used for verification.

CHAPTER THIRTEEN

Spoken Help for a Car Stereo: An Exploratory Study

GOVERT DE VRIES

Kodak Imagination Works, 490 California Street,

Suite 300, Palo Alto, CA 94306, USA

and

GRAHAM I. JOHNSON

NCR, Technology Development, Cognitive Engineering, Kingsway

West, Dundee DD2 3XX, Scotland

ABSTRACT

This chapter reports an investigation of the use of spoken (audible) help in an attempt to increase the usability and learnability of a high-end (i.e. relatively expensive, multi-featured) car stereo. To investigate the usefulness of audible spoken help for a car stereo, a simulation was constructed within which experimental help systems were incorporated. To validate the simulation, the 'real' car stereo (an existing consumer product, acting as the control condition) was compared empirically with its simulation.

An experiment, using the four conditions of actual product, simulation of actual product, and two simulations employing different spoken help functions, was undertaken. Forty participants in a between-subjects design carried out specified tasks with the car stereo version, according to experimental conditions. User performance, specifically task completion and number of button presses, and subjective reactions were measured. The simulated spoken help versions clearly assisted users with their tasks, and in general were responded to favourably when compared to performance with the control simulation and its product equivalent.

This exploratory study provided further insight into users' concerns about car stereo usability and the overall results revealed the potential of spoken help facilities for novice users. We can conclude that the use of audible, spoken help, whether as a global option or as a specific 'button help', enhances the performance of, and is acceptable to, novice users of this consumer product type. Finally, this investigation also demonstrated the use of a (Hypercard) simulation versus the 'real' counterpart (the simulated car stereo) to be generally valid and appropriate for this type of evaluation. In order to progress this direction, examination of the design of spoken help dialogues, user control of these, and issues concerning 'longitudinal' use of the product type are recommended.

13.1 INTRODUCTION

As technology becomes less expensive, today's professional products easily become tomorrow's consumer products. Consumer products currently available have so many functions and are introduced in such a short period of time, that the usability of some of these products has been noted as a cause for concern (Norman 1988).

A contemporary in-car system is a good example of a consumer product within which the functionality has recently become considerably more complex. As a result of increased computerisation and miniaturisation, most car stereos currently available comprise both a greater number of, and a more complex arrangement of, functions than professional audio equipment did five years ago. User-centred studies have demonstrated, predictably, that users do not always take full advantage of the functionality offered by their car stereo system (van Nes and van Itegem 1989).

Current trends mean that user interfaces to in-car equipment are likely to become more complex, as the general aim is to develop comprehensive, multi-functional in-car systems. In order to prevent the introduction of advanced systems which severely impede the user's primary task, that of driving, considerable attention needs to be given to all usability aspects of these products (Johnson and Rakers 1990).

The study described here investigated whether the use of spoken help can increase the usability, and more specifically the learnability, of a high-end (i.e. relatively expensive) multi-featured car stereo. Intuitively, the use of audible help would seem advantageous
(over other forms) since it is less likely to conflict with the chiefly visual attention demands of the primary (driving) task.

Because of the simulation approach taken, we had to pose the question: "How and in what way does the use of a computer simulation influence user performance and perception compared to user performance and perception with a 'real' car stereo?" It is apparent that a great many usability studies and product development activities make numerous assumptions about the general validity of computer-based simulations.

Our main research questions were:

- Is a car stereo equipped with the audible help system more learnable and usable than the same system without an audible help system?
- What are the strengths and weaknesses of the help system and what help should be given in what context?

13.2 SPEECH AS AN INFORMATION DISPLAY MEDIUM

A person can listen to spoken instructions and simultaneously drive a car, as evidenced in the case of driving tuition. The desirability of using vocal instuctions for real-life navigation tasks is clear, given that high attentional loading of the visual channel is considered to be a major cause of accidents (Weirwille *et al.* 1988; Rockwell 1988; Swift and Freeman 1986).

However, there are two important disadvantages of spoken output. Firstly, comprehension of spoken language is slower than reading. Secondly, spoken language has the disadvantage of being transitory. Unlike written information it cannot be easily referred to after the event.

There are also a number of practical advantages of speech in comparison with visual information display. Speech can operate over distance, makes use of the universality of the spoken language and allows processing in other modalities. Another advantage of speech as a means of instruction provision is that it can be incorporated into a system itself, and therefore increases the chance that the user will use the help part of that system. However, it is important to note that inclusion does not necessarily ensure reference.

13.3 THE CAR STEREO — PHILIPS DC794

The car stereo (see Figure 13.1) was selected because it was regarded as one of the top models ('high-end') in the range of car stereo products, and offers a broad functionality. This model provides FM, LW and MW radio combined with a sophisticated auto reverse cassette player. Its user interface falls within the standard (DIN, within Europe) area for the control layout.

The DC794 offers a grand total of thirty-nine discrete basic functions. As seen in Figure 13.1, the layout of controls is relatively conventional in that a central display and cassette player is bordered by push buttons of slightly different types.

Figure 13.1 The Philips DC794 car stereo.

13.4 DEVELOPING THE PRODUCT SIMULATION

The development of a simulator is not, or should not be, simply a matter of requiring engineers to reproduce an operational system. Development of a simulator begins with an analysis of what one wishes the simulator to do. This includes an analysis of the system's purpose, its tasks and, most important of all, which part of the 'real' system is of interest.

The simulation of the car stereo was built in Hypercard (see Figure 13.2). Complete functionality was implemented in the simulation, except for the remote control, the security code and one of the two possible methods to programme a preset station. Different fragments of sounds (mostly music) were recorded and implemented in the simulation. When a new station was chosen, or when, for instance, the cassette was wound to the next track/song, a pre-recorded music fragment was produced, lasting about five seconds.

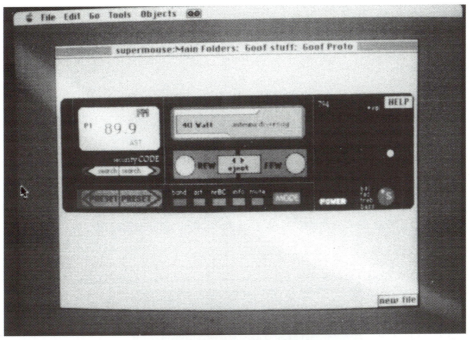

Figure 13.2 Simulation of car stereo.

All user actions (such as button presses, button times and task times) made during the experimental trials were logged within the constructed simulation and written to a data file which was later used for statistical analysis.

13.5 CONSTRUCTING THE AUDIBLE HELP SYSTEMS

The construction of the audible help, its structure and content, made use of the results of:

- a GOMS analysis (Card, Moran and Newell 1983) of car stereo use, and
- the results of a usability evaluation based on the '20 Questions Method' (Grover 1983), and relevant literature consulted. The results of these studies are fully reported in de Vries and Johnson (1992).

On the basis of these analyses, two audible help systems were constructed:

1. A help button with which the user can request basic information about specific buttons.
2. A system which automatically provides help; the user can switch this type of audible help 'on' or 'off'.

The help systems are described in Figures 13.3 and 13.4.

BUTTON-HELP SYSTEM
a scenario of a user who asks help about the 'mode button'

1.   - User presses the Help Button.

- Help system gives user information about how he/she can get help.
"You can now ask help about a button by pressing it."

- -

2.   - User presses the button he/she wants information about.

- System gives user information about the button the user pressed.
"Mode. When you press this button, the system switches over from radio to cassette or from cassette to radio.
You can only change from radio to cassette when a cassette is in the system."

- -

3.   - System returns to original mode.

Figure 13.3 Button-help system.

Feedback & Guidance SYSTEM
a scenario of a user who gets help about the 'mode button'

The system is in 'radio' mode, no cassette is in the system.

1.   - User presses the Help Button.

    - Help system gives user information about the feedback and guidance system.
"*The help system is now on. When you press a button, you get information about that button. By pressing the help button again, the help system is switched off.*

- -

2.   - User presses the button he/she wants to operate.

    - System gives user information about the button user's action.
"*It is not possible to change from radio to cassette because at present no cassette is in the system.*"

3.   - System returns to original mode.
"*The help system is now switched off. By pressing the help button again, the help system is switched on.*"

N.B. Note that the feedback & guidance system stays active until the user switches it off by pressing the help button. When the feedback & guidance system has been activated, the user automatically gets information about all of the buttons he/she presses (except for the search button and the volume control).

Figure 13.4 Feedback and guidance system.

13.6 THE EVALUATION METHOD

Design

Four groups of ten subjects were used in a between-subjects design. One group carried out their tasks using the 'real stereo' condition with the Philips DC794 car stereo, another with the simulation not equipped with a help system. These formed the two control conditions. The experimental conditions made use of the simulation equipped with (a) the 'button help system', and (b) the 'feedback and guidance system', respectively. These conditions are shown in Table 13.1.

A between-subjects design was chosen because of the obvious issues of learning effects (all subjects were initially naive with respect to the car stereo) and because of the practicalities of experimental time per subject.

Comparisons between the subjects' performances with the stereo were made, using the number of tasks successfully accomplished as the main dependent measure.

Table 13.1 Overview of groups in the experiment.

Group Treatments	Number of Subjects
Group 1: 'real-radio' (Philips DC794) CONTROL	10
Group 2: computer simulation of car stereo DC749 CONTROL	10

Group 3: computer simulation of car stereo DC749
            with audible button help
            EXPERIMENTAL
	10

Group 4: computer simulation of car stereo DC794
            with audible feedback & guidance
            EXPERIMENTAL
	10

Subjects

Subjects were screened according to pre-defined criteria, with the intention of keeping subject variability to a minimum. The most important criterion was never to have owned, or to have frequently used, a car stereo, which had the same or similar functionality as the Philips DC794.

For the purpose of the experiment the subjects were car stereo novices. They were placed in a situation of learning to operate the car stereo without the use of a manual, in a non-driving situation, in many ways equivalent to:

- People first using their car stereo in a non-driving situation, immediately after the car stereo has been fitted into the car.
- In practice manuals get lost and will often not be available in, for example, second-hand cars with built-in car stereos.
- In rental cars a manual is often not available, and the user is often not familiar with the operation of the car stereo.
- People often choose to experiment on a trial and error basis with products such as car stereos.

Apparatus

Philips Design's usability laboratory was used for the experiment. All trials were recorded using a video camera. Sound was also recorded. The camera was positioned so that it was possible to see which buttons were pressed. The 'real stereo' condition and performed tasks used the Philips DC794 which was mounted in a 'point of sale' unit (see Figure 13.5). Such a unit is often found where car stereos are on display so that the public can try them out.

Figure 13.5 Point-of-sale unit incorporating Philips DC794 car stereo.

Participants in the 'simulation conditions' (see Figure 13.6) performed tasks using the Hypercard computer simulation of the Philips DC794.

Figure 13.6 HyperCard simulation conditions.

All button presses during task performance were recorded and written to a log file.

Our study investigated the usability and, more specifically, the learning process of using the car stereo. Therefore, we assumed that people would try out and explore the car stereo for some time while not driving.

Tasks

Thirteen separate tasks were chosen with the intention of covering as much of the stereo's functionality as possible, as well as covering a range of difficulty. The tasks also had to follow in a logical and realistic sequence. Some examples of the tasks given are — Switch power on; Tune to a music station; Program a station; Change to cassette mode.

Measure

Task Measures:
Effectiveness (number of tasks successfully completed) and Efficiency (number of actions and time taken to accomplish task(s)).

Subjective Measures:
To obtain an overall impression of user satisfaction we used three elements:
(i)   General questionnaire — Eighteen questions relating to the stereo's usability, format based on a questionnaire developed by Chin, Diehl and Norman (1988). An example of a question is:

To learn to operate the system is:

Very difficult								Very easy	
0	1	2	3	4	5	6	7	8	9

(ii)   Problem ratings, a questionnaire rating the seriousness (or triviality) of some of the issues they may have encountered while using the stereo, based on an adaptation of section 10 of the evaluation checklist developed by Ravden and Johnson (1989). Eight questions were asked. An example of a question is:

Knowing what the result of an action is:

No problem			Major problem
1	2	3	4

(iii) Semi-structured interview — open-ended questions to gather opinions of the stereo, format of this based on Ravden and Johnson (1989). In the semi-structured interview six questions were asked. An example of a question is: What was the worst or most irritating aspect in the operation of the car stereo?

Procedure

The subject was shown the usability laboratory (experimental room) and camera equipment. After this, a verbal introduction was given on the purpose of the experiment.

Subjects were then asked to complete a consent form and a questionnaire to record relevant personal details.

Subjects who participated in one of three simulation conditions were then asked to perform two pre-tests to familiarise them with the use of both mouse and computer.

The experimenter read aloud the tasks and the subjects performed them with the stereo, or with the simulation of the stereo (depending on the condition). Finally, participants were asked to fill in the usability and problem questionnaire and the interview was held.

13.7 RESULTS AND DISCUSSION

How good was the simulation?

Each subject was set the same thirteen tasks in the same order. A Kruskal-Wallis showed that there was a significant difference for task accomplishment between the four conditions (p = .001). This result does not appear to be due to a difference between the 'real' car stereo and the 'simulation' condition, because no significant difference was found from an unpaired t-test for task accomplishment between these conditions (p = .86). Mean number of accomplished tasks in the 'real' car stereo and the 'simulation' condition was exactly the same, with an average of 46% of tasks being successfully accomplished.

A significant difference was found for the scores on the usability questionnaire between the four conditions (p = .021, 1-way ANOVA). Between the 'real' car stereo (mean = 4.0) and the 'simulation' condition (mean = 4.8) no significant difference was found according to an unpaired t-test for scores on the usability questionnaire between these conditions (p = .09).

All subjects were also asked to complete a questionnaire rating the seriousness (or triviality) of some of the problems they had encountered while using the stereo. A significant difference was found for the scores on the problem questionnaire between the four conditions (p = .04, 1-way ANOVA). This result does not appear to be due to a difference between the 'real' car stereo (mean = 2.6) and the 'simulation' condition (mean = 2.3) because no significant difference was found, according to an unpaired t-test for scores on the problem questionnaire, between these conditions (p = .26).

Remarks made by subjects in the 'real' and the 'simulation' condition were very much alike. Furthermore, subjects in both conditions encountered the same kind of problems in their attempts to accomplish the tasks, and applied the same sort of strategies to discover how tasks should be accomplished.

The interview results also indicated no differences between the two control conditions. Subjects in both the 'real' and the 'simulation' condition indicated that they encountered most problems while performing the programming, rewinding to the beginning of the tape and winding the tape forward to the next song/track.

It can be concluded that the quantitative and qualitative data clearly indicate that both the realism (fidelity) and the comprehensiveness of the simulation used in this experiment were satisfactory.

Did audible help improve the usability of the car stereo?

Since the results indicated that the fidelity of the car stereo simulation was good, it was assumed that the simulations equipped with spoken help systems were a valid means of assessing the usability of this car system.

Subjects in the 'button-help' condition and the 'feedback and guidance' condition accomplished more tasks that subjects in the 'simulation' condition, and in the 'real' car stereo condition (see Figures 13.7 and 13.8). A significant difference between the 'simulation' condition and the 'button help' condition was found according to an unpaired t-test for accomplished tasks (p = .0002). Also, a significant difference between the 'simulation' condition and the 'feedback and guidance' condition was found according to an unpaired t-test for accomplished tasks (p = .01).

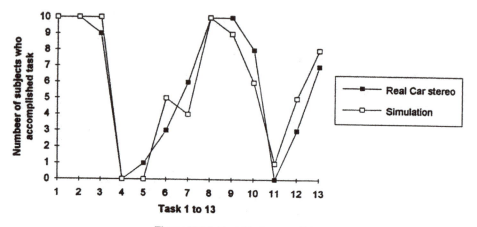

Figure 13.7 Subjects' task accomplishment 1.

Subjects rated the 'button-help' system and the 'feedback and guidance' system as clearly more usable than subjects in the simulation condition and in the 'real' care stereo condition. A significant difference between the 'simulation' condition and the 'button-help' condition was found for the score on the usability questionnaire according to an unpaired t=test (p = .005). No significant difference was found between the 'simulation' condition and the 'feedback and guidance' condition for the mean score on the usability questionnaire according to an unpaired t-test (p = .20).

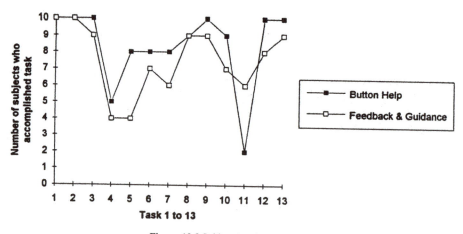

Figure 13.8 Subjects' task accomplishment 2.

Subjects reported in the 'feedback and guidance' condition that they were annoyed by the same help messages being repeatedly given. Some subjects used the 'feedback and guidance' system as if it were a 'button-help' system by switching it off after each help request.

Subjects rated the 'button-help' system and the 'feedback and guidance' system as causing less usability problems than subjects in the 'simulation' condition and in the 'real' car stereo condition.

A significant difference between the 'simulation' condition and the 'button help' condition was found for the score on the problem questionnaire according to an unpaired t-test (p = .005).

No significant difference between the 'simulation' condition and the 'feedback and guidance' condition was found for the mean score on the problem questionnaire according to an unpaired t-test. It may be that subjects in the 'feedback and guidance' condition rated the system as causing the same level of problems because they became annoyed by the fact that they were given the same help messages repeatedly.

All subjects said that without the help system they would have accomplished fewer tasks, and both in the 'button-help' and the 'feedback and guidance' condition the audible help was rated as very useful.

Evaluating the help systems: What help should be given in what situation?

Both help systems were constructed as the most extreme examples of their class. Although it was predicted beforehand that the help would differ from function to function, no attempt was made to construct a help system, which was a combination of a 'button-help' and a 'feedback and guidance' system. It was felt that the strengths and the weaknesses of each system should be evaluated.

Some people used the 'feedback and guidance' system as if it were a 'button-help' by switching it off after each help request. Some subjects became annoyed when they had to listen repeatedly to the same message, as was the case with the preset function. It is therefore recommended that no help is given by the 'feedback and guidance' system for buttons which are likely to be pressed frequently. Another solution to this problem is that the stereo stops giving automatic 'feedback and guidance' after the same help message has been given a certain number of times within a set period.

It remains debatable whether in some cases the system should give audible feedback automatically, e.g. when the user presses the mode button while no cassette is in the system. This should only be done to give the user feedback about errors, not during error-free interaction. However, it should always be possible to switch this off permanently.

Limitation and caveats

Simulation with help systems

Although this study clearly indicated that the simulation fidelity was high, some questions with respect to the generalisability of the results with the spoken help system remain. Subjects' reactions might have been positively biased because the help was implemented on a computer simulation and might have been less favourable had the help system been built in a real car stereo. Further research should be carried out to investigate this issue of acceptability.

Stationary car stereo use

This study investigated the usability of car stereos in a stationary, non-driving condition. Firstly, it can be assumed that most users will 'play' with and explore the product immediately after it has been installed, therefore most likely in a non-driving situation.

Secondly, if users benefit from an audible spoken help system in a stationary situation, it is expected that they will benefit even more from this help system while driving.

It may be expected that an audible spoken help system will be much more frequently used than a paper-based user manual. Westendorp, van Weezel and Moraal (1993) compared an on-line help system and a user manual. Results of this study were that all people used the on-line help system while none of the subjects used the user manual. It seems that consulting an on-line help system is perceived as less deviating from the task than consulting a user manual. Given the fact that people are very task oriented (e.g. Carroll and Rosson 1987), having a built-in help system seems to increase the chances that it will be used.

Effectiveness and efficiency
The results indicate that users in the help conditions did not perform the tasks with less button presses than subjects who used the simulation without help. One might have expected that people in the help conditions would not only be more successful but also more efficient in performing the tasks that were set. Users in the help conditions might have been less efficient because they effectively had an additional task — operating the help system.

13.8 REFERENCES

Card, S.K., Moran, T.P. and Newell, A., 1983. *The Psychology of Human Computer Interaction.* Lawrence Erlbaum Associates, Hillsdale, New Jersey.

Carroll, J.M. and Rosson, M.B., 1987. Paradox of the active user. In *Interfacing Thought: Cognitive Aspects of Human Computer Interaction,* J.M. Carroll (Ed.), MIT Press, Cambridge, Mass.

Chin, J.P., Diehl, V.A. and Norman, K.L., 1988. Development of an instrument measuring user satisfaction of the human computer interface. In *Proceedings of the CHI'88: Human Factors in Computing Systems,* May 15-19, Washington DC, ACM Press, N.Y., pp. 213-218.

Grover, M.D., 1983. A pragmatic knowledge acquisition methodology. In *Proceedings of the 8th International Joint Conference on Artificial Intelligence,* IJCAI-83.

Johnson, G.I and Rakers, G., 1990. Ergonomics in the development of driver support systems: putting the user back in the driving seat! In *Proceedings of the 22nd International Symposium on Automotive Technology and Automation (ISATA 22),* 14-18 May, Florence, Italy, pp. 336-342.

Nes, F.L. van and Itegem, J., 1989. *Usage of car-radio controls.* Institute for Perception Research (IPO), Eindhoven University of Technology, Report No 691.

Norman, D.A., 1988. *The Psychology of Everyday Things.* Basic Books, N.Y.

Ravden, S.J. and Johnson, G.I., 1989. *Evaluating Usability of Human Computer Interfaces: A Practical Method.* Ellis Horwood, Chichester.

Rockwell, T.H., 1988. Spare visual capacity in driving-revisited. New empirical results for an old idea. In *Vision in Vehicles — II,* A.G. Gale, M.H. Freeman, C.M. Haslegrave, P. Smith and S.P. Taylor (Eds), pp. 317-324.

Swift, D.W. and Freeman, M.H., 1986. The application of head-up displays to cars. In *Vision in Vehicles, Proceedings of the Conference on Vision in Vehicles,* A.G. Gale, M.H. Freeman, C.M. Haslegrave, P. Smith and S.P. Taylor (Eds), North Holland, Amsterdam, pp. 249-255.

Vries, G. de and Johnson, G.I., 1992. The use of GOMS and the 'limited information task' for the evaluation of a car radio system. *Ergonomics,* **17**, 6, pp. 2-9.

Westendorp, P.H., Weezel, K. van and Moraal, J., 1993. On-line help or hard copy help for a photocopier. *Ergonomics*, **18**, 3, pp. 2-6.

Weirwille, W.W., Antin, J.F., Dingus, T.A. and Hulse, M.C., 1988. Visual attentional demand of an in-car navigation display. In *Vision in Vehicles — II*, A.G. Gale, M.H. Freeman, C.M. Haslegrave, P. Smith and S.P. Taylor (Eds), pp. 317-324.

CHAPTER FOURTEEN

Prototypes on Trial

M. J. (THEO) ROODEN

Faculty of Industrial Design Engineering, Delft University of Technology,

Jaffalaan 9, 2628 BX Delft, The Netherlands

14.1 INTRODUCTION

In Chapter 12 on usability testing under time-pressure, Kahmann and Henze make the point that, while rapid trialling techniques are necessary in industry, the comparison of different techniques and detailed observation is the domain of academe. In a design process, anticipation of future usage is enhanced by users' trialling with prototypes of the intended product, because the product is not yet available. These prototypes range from rough sketches to working prototypes. A prototype is defined by Ulrich and Eppinger (1995) as 'an approximation of the product along one or more dimensions of interest'. Kaulio (1997) asserts that this definition of a prototype is used by researchers and includes all types of representations of the product during development. In our research we have adopted this broad definition. The research focuses on users' trialling with prototypes. One question to be answered is which differences and resemblances there are between the use of a product, and derived design models, in terms of perception, cognition and use actions. The origins of observed differences can then be investigated, leading to proposals for how to execute users' trials during design processes, and how to construct prototypes for users' trialling as a design tool. Although one usually wants to represent a high level of detail of the design in a prototype, limited time and cost often prevents this from happening. Certain aspects of the design will not be realised in the prototypes, with possible consequences for the generalisability of the results from users' trials with these prototypes to usage of the intended product. A series of experiments has been planned to attempt to gain insight into these matters, and one of the early experiments is discussed in this chapter, leading to suggestions for the improvement of users' trialling with prototypes.

14.2 THE STUDY-SET UP

Product

A series of users' trials comparing prototypes and a real product was carried out. The product chosen was a blood pressure monitor to be used at home (see Figure 14.1). Usage of such a product involves manipulation, force exertion, operation of controls and interpretation of feedback/feedforward on an LCD-display, making it possible to study a wide range of user activities. To measure one's blood pressure, the pump and the cuff have to be connected to the monitor, the cuff has to be applied around the left upper arm, and secured with Velcro adhesive fabric. Then the monitor has to be switched on, and the cuff inflated by hand to a level above the expected systolic pressure. The measurement then takes place during which the cuff is deflated automatically. After the results (diastolic and systolic blood pressure and pulse) have appeared in the display (LCD), the rest of the air can be released by pressing the valve on the pump. The cuff can then be removed, the parts disconnected and the monitor switched off.

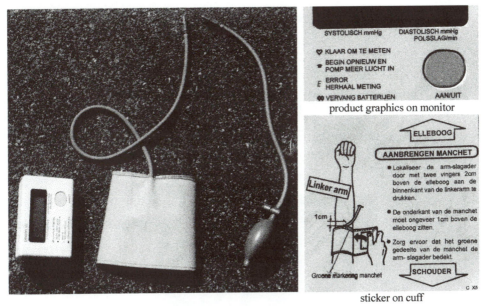

Figure 14.1 The blood pressure monitor.

Prototypes

Prototypes were constructed retrospectively to accommodate comparisons of operations by users. The prototypes were a set of drawings, a mock-up (see Figure 14.2), and a non-functioning product which was the product with the batteries removed and with prevention of air flow from pump to cuff. Information on the display was simulated with cards in all prototypes. These cards were put in place at the appropriate moments by the test leader. We let the prototypes differ in certain characteristics on which we wanted to study the 'influence' on user activities. For instance, the mock-up and the drawings mainly differed in manipulability. The third prototype and the real product mainly differed in functioning. The two 'primitive' prototypes lacked many use cues (see Table 14.1). This variation made the study of these characteristics (functioning, manipulability, and the role of use cues) possible. However, at the same time it limited generalisability because such prototypes are not realistic in a design process.

Procedure

Subjects (n=36) were invited into our laboratory and asked to operate the various prototypes and the real product. There were 63 trials. Three x nine (=27) subjects went through two trials, first operating one of the prototypes, and then using the real product. One group of nine subjects only used the real blood pressure monitor. Subjects were casually requested to verbalise their perceptions and use actions concurrently. Afterwards the video-taped operations were reviewed with the subjects and they were retrospectively probed for further information on perception, cognition and use actions. Also some questions about relevant experience and familiarity with (parts of) the product were asked. From the videotapes, lists of use actions and the difficulties experienced were gathered. Information from the verbal reports was further used to locate origins of certain use actions.

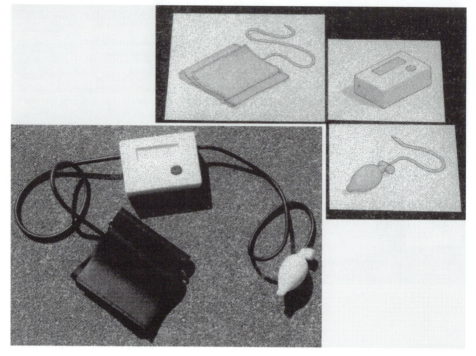

Figure 14.2 Prototypes of the blood pressure monitor: a mock-up and a set of drawings.

14.3 RESEMBLANCES AND DIFFERENCES BETWEEN OPERATING PROTOTYPES AND REAL PRODUCTS

In this section operation of the prototypes and usage of the blood pressure monitor are compared with regard to use actions and interaction difficulties, both in nature and number. Because identifying interaction difficulties is a major goal of users' trialling, this is discussed first.

Interaction difficulties

Interaction difficulties were identified by viewing the video-tapes of usage. Actions or intentions for actions leading to failure (i.e. not getting results) and verbal expressions of non-comprehension were identified as interaction difficulties. A total of 94 distinct difficulties were revealed in the 63 trials. The resolution of description of the interaction difficulties was set at a level where the information was supposed to be of help in generating design solutions, which means that they were described at a rather detailed level. For instance 'a subject wants to connect the plugs of the pump and the cuff to each other' and 'a subject wants to connect the pump directly to the cuff' were regarded as distinct difficulties, since they would probably call for different design solutions. Usage of (only) the real blood pressure monitor showed twenty distinct interaction difficulties. Only some of these difficulties were revealed by operation of the prototypes, while operation of the prototypes generated a number of extra difficulties (see Figure 14.3 and Table 14.2). This shows that although the *number* of difficulties between prototypes and real product may correspond, the *nature* of these difficulties may differ.

Table 14.1 Representation of use cues in the prototypes.

yes = represented no = not represented	drawing	foam model	non-functioning product	product
monitor				
display	yes	yes	yes	yes
information on the display	partly[1]	partly[1]	partly[1]	yes
graphics (text and symbols)	no	no	yes	yes
on/off button	yes	yes	yes	yes
auditory signals	no	no	no	yes
sockets	partly[2]	yes	yes	yes
pump with hose				
familiar form pump	yes	yes	yes	yes
familiar form plug	yes	yes	yes	yes
cues in functioning	no	no	no	yes
cuff with hose				
familiar form cuff	yes	yes	yes	yes
Familiar form plug	yes	yes	yes	yes
Sticker with instructions	no	no	yes	yes
green strip to indicate orientation	no	no	yes	yes
Velcro fastening	no	no	yes	yes
pressure while functioning	no	no	no	yes

[1] Information on the display was presented in discrete steps, symbols did not flash.
[2] Only one socket was visible in the drawing.

To improve the design, the designer needs to know which difficulties encountered in operation of prototypes are expected to occur with the real product. It remains to be researched whether an 'expert' can predict which of the difficulties with the prototypes will also occur with the real thing, and whether he or she can anticipate some of the 'real' difficulties which were not revealed in operating the prototypes. Not including such predictions, and the fact that the prototypes were not 'design-process-realistic' has probably blackened the picture in this experiment. Whether early users' trialling is more useful than would be concluded from Figure 14.3 and Table 14.2 remains in question.

There were six interaction difficulties, which occurred more than ten times. These are presented in Table 14.3. Origins of difficulties in certain conditions are discussed later.

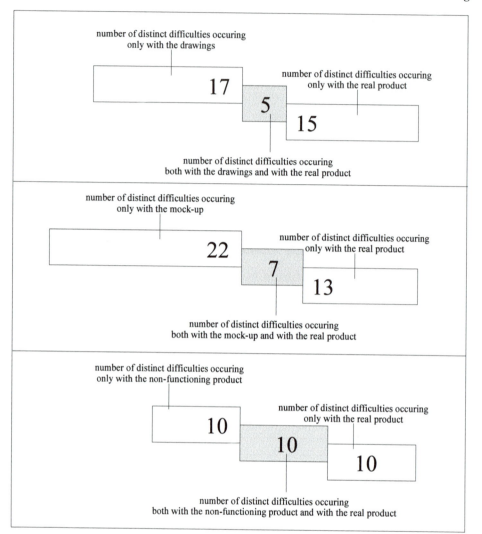

number of distinct difficulties occuring
only with the drawings

17

number of distinct difficulties occuring
only with the real product

5

15

number of distinct difficulties occuring
both with the drawings and with the real product

number of distinct difficulties occuring
only with the mock-up

22

number of distinct difficulties occuring
only with the real product

7

13

number of distinct difficulties occuring
both with the mock-up and with the real product

number of distinct difficulties occuring
only with the non-functioning product

10

number of distinct difficulties occuring
only with the real product

10

10

number of distinct difficulties occuring
both with the non-functioning product and with the real product

Figure 14.3 Difficulties experienced with the prototypes and with the real product.

Table 14.2 Resemblances in interaction difficulties between each of the prototypes and the real product. (see also Figure 14.3).

20	distinct difficulties when using the real product (n=9)[1]
8 of these 20	difficulties were observed only in operation of the real product (not shown in Figure 14.3)
5 of these 20	difficulties were also observed in operation of the drawings (n=9)
7 of these 20	difficulties were also observed in operation of the mock-up (n=9)
10 of these 20	difficulties were also observed in operation of the non-functioning product (n=9)

[1] These subjects did not operate a prototype of the blood pressure monitor.

Table 14.3 Most frequently occurring interaction difficulties and possible origins of these difficulties.

	drawings	mock-up	non-functionin g product	only product	product after having operated a prototype
	(n=9)	(n=9)	(n=9)	(n=9)	(n=27)
a subject inflates the cuff without switching the monitor on first	6[1]	4	0	1	2
This difficulty happens almost exclusively with the primitive prototypes. One reason is that the on/off switch is then not recognized as such. Another reason may be that switching the monitor on is not necessary when telling how to operate the blood-pressure monitor. With the real thing it is necessary to switch the monitor on to get results.					
a subject wants to connect the pump directly to the cuff	8	2	1	0	2
This is a typical 'drawings-difficulty'. Only one socket in the monitor is visible on the drawings. Trying to connect the pump to the cuff is logical in a way, because the cuff is to be inflated. Subjects may have the same intention with the other prototypes and the real product, but they can then find two sockets in the monitor.					
a subject does not understand the results of measuring the pulse	4	1	2	3	6
Operation of the prototypes resembles usage of the real product.					
a subject inflates the cuff without waiting for the signal to start	0	0	2	3	6
This difficulty occurs most often when using the real product. One of the reasons is that less subjects get to this stage in the prototype-conditions. Furthermore, this is a difficulty arising from the functioning of the monitor. The simulation of the start-up phase of the interaction triggers a different response. Subjects waited patiently for the test-leader to put the cards on the display.					
a subject wants to press the air release valve after inflating the cuff	1	3	2	2	3
Operation of the prototypes resembles usage of the real product.					
a subject does not inflate the cuff to a height above the blood pressure	1	1	0	4	13
This occurs almost exclusively when using the real product; a difficulty arising in the functioning of the product.					

[1] Number of subjects experiencing the mentioned interaction difficulty.

Use actions

Another aspect on which resemblances and differences between operating prototypes and using a real product can be studied is the number of use actions. It turned out that with the real product more actions were performed (see Table 14.4). Explanations are that less actions are possible with the primitive prototypes (for instance reading the sticker with instructions on the cuff), and that the real product gives opportunities for learning by trial and error, which can motivate users to carry on. Perhaps subjects are also better motivated when they really can get results.

The extra actions performed with the real product were mainly switching the monitor on and off, and pumping without success. With the models, inflating to the desired level merely involved showing the method of squeezing, and knowing the level required. With the real product, the speed of pumping also mattered, and the building up of pressure in the arm made subjects have doubts about the required level of inflation. They needed more attempts to inflate the cuff to the desired level.

Table 14.4 Number of actions (median per condition).

	Number of actions
Drawings (n=9)	12
Mock-up (n=9)	11
Non-functioning product (n=9)	15
Product (n=9)	27
Product after having operated a prototype (n=27)	21

14.4 ORIGIN OF DIFFERENCES IN OPERATION BETWEEN PROTOTYPES AND REAL PRODUCTS

In the previous section differences between operation of prototypes and usage of the real product were identified. Some origins of these differences were already mentioned when the more frequently occurring difficulties were presented (see Table 14.3). In this section an overview is given of possible origins of the differences. Also, some suggestions are presented (in italics) to improve primitive prototypes and the methods of users' trialling with such prototypes.

Elicitation of use actions

'Operation' of a drawing differs from usage of a real product, because use actions can not actually be performed, but have to be verbalised. With a drawing subjects tell what they would do, and it is for them to decide what they find relevant to mention. This may differ from what the researcher would like to have mentioned. Spoken information is less refined than displayed usage with a real product, or with manipulable prototypes. Certain use actions proved difficult to verbalise. For instance subjects are not used to talking about manipulations and do not have an adequate vocabulary. This became apparent when subjects had to talk about applying the cuff. No details were mentioned spontaneously by the subjects.

In addition to verbal information, subjects sometimes pointed at parts of the drawing or at their body to illustrate what they meant. Subjects often gestured when talking about applying the cuff. However, some subjects did not touch the drawings at all.

Although presented in an awkward orientation, the drawings were not re-arranged by those subjects.

When subjects operate a drawing, and use actions have to verbalised, an active role by the test-leader is desirable. He or she should probe for specific information and details during the operation. Additional means of expression, such as gesturing and pointing, should be encouraged.

Visibility of information

In a drawing some sides of the design may not be represented, as was the case in the experiment. These invisible sides can also contain 'information'. Subjects can not make use of this non-represented information. For instance, most subjects did not identify a second socket in the other side of the drawn monitor, which is the case with the real monitor. On the other hand, some subjects can project expected or desired 'information' on to the invisible sides. For example those subjects who assumed a socket on the other side of the cuff to which to attach the pump, which was apparently more logical to them than a socket in the monitor. Initially subjects expected all relevant information to be represented in the drawings. When they missed information they projected some of this information onto invisible sides of the product.

Try to show all sides of the design in a drawings-prototype, for instance by presenting a set of drawings, or a computer animation. However, it must be realised that certain subjects are not familiar with 'reading' drawings. Visual literacy is often a basic assumption of designers. Hidden information should be communicated to the subjects.

Primitive prototypes leave room for alternative interpretation

As already mentioned, in the drawing the pump can be 'connected' to the cuff. This actually happened in the users' trials. Without the intervention of the test-leader, this 'mistake' only becomes apparent when the subjects start pumping and the figures on the display do not rise. (The figures in the LCD show the air pressure in the cuff at that moment.) Some subjects, however, do not expect these figures to be simulated with cards on the display, and they stick to the idea that the pump can be directly connected to the cuff.

After having inflated the cuff, it is deflated automatically. Subjects who do not expect this sometimes try to deflate the cuff manually by turning the air release valve on the pump. They then guess that the deflation which follows is caused by turning the valve. However, the air release valve is a press button, which releases the remaining air after the measurement is completed. Turning the valve has no effect; the cuff deflates automatically.

When subjects mention that they expect the real product to differ from the prototypes, by adding or changing certain characteristics, they should be corrected.

The subjects should be told at an appropriate moment what is simulated by what in the prototype. For instance, when subjects have identified the LCD, they can be told that information on the display will be simulated with cards.

Altered appearance

A lot of use actions change the product's appearance. For instance, connecting the pump and the cuff to the monitor, and applying the cuff around the arm. When these changes are not made visible in the prototypes as is the case in conventional drawings, subjects may forget to execute certain use actions, such as removing the cuff from their arm and disconnecting the parts, as happened in the experiment.

Make foreseen changes in the appearance visible in the prototype by preparing some extra drawings.

Leaving out product graphics

In the primitive prototypes (the drawings and the mock-up) product graphics were left out. Sometimes this made a difference. For instance, absence of the text 'on/off' adjacent to the on/off-button made identifying this button more difficult. Absence of the rest of the product graphics did not seem to matter, partly because of the poor quality of the graphics on the product, which gave confusing information. In a design process the impact of product graphics can only be tested when these graphics are represented in the prototype.

Represent product graphics in the prototypes as far as designed at the moment of trialling.

Representation of physical aspects

Some physical characteristics need not be represented precisely, because they are obvious. For instance, the pump did not need to be squeezable for subjects to understand that squeezing was the way to operate the pump; there simply is no alternative. However, the fact that the Velcro fastening was left out, made subjects search for various fixation mechanisms.

Physical aspects can be left out when there is one obvious way of operation, but this is not often clear beforehand.

Functioning or not

Users' goals seem to be different when operating non-functioning prototypes compared to using a real blood pressure monitor. In the last case it is all about getting results. The user cannot skip a step in the interaction; also trivial and straightforward actions must be performed. When operating a non-functioning prototype, subjects seem to be explaining what they would do if it were a real product. When telling and showing this, trivial actions may be omitted since they are considered too obvious or unimportant to be mentioned. This may have been the cause of a number of subjects not mentioning the need to switch the monitor on. This action may be so logical that the subject may think 'the test leader will assume that I would do that.' Also, the order of actions is of less importance to the subjects, when getting results is not the primary goal. It makes no difference to the result if they demonstrate how to pump, even when the connections are not made.

Functioning prototypes are preferred to non-functioning ones. When applying non-functioning prototypes it should be realised that users' goals may be different, and that certain use actions may not be mentioned, or that the order of actions may be inaccurate.

Some interaction difficulties are only revealed when a prototype or product is functioning. In our experiment, with the non-functioning prototypes more subjects succeeded in getting results than with the real product. For instance, air pressure building up in the cuff, which is felt in the arm, had an effect on subjects' actions.

Add users' trials with existing 'similar' functioning products to the design process to anticipate interaction difficulties, connected to the functioning of the product.

Response times

Simulation of information from the LCD differed from the feedback from the real product. When subjects were talking the test leader sometimes waited for them to finish a sentence, before putting on the next display. This is probably due to social conventions.

Simulated feedback should be supplied by an anonymous person, not by the test leader. This makes it easier to stick to the response times of the real product. This may also be facilitated by a computer simulation.

14.5 METHODOLOGICAL ASPECTS OF USERS' TRIALLING WITH PROTOTYPES

A range of differences between operation of prototypes and usage of the real product could not be related to differences in characteristics between the prototypes and the real blood pressure monitor. For some differences such a relation may still exist. Information from subjects was not as helpful in identifying origins of differences as was hoped for. It proved difficult to elicit users' perception and cognition, both during the operation and afterwards. There was also a number of artifacts in the research which blurred the results. These two topics are discussed below, because they are deemed to be general issues in users' trialling.

Eliciting perception and cognition

When it is known what users perceive and think during product usage (or operation of prototypes), the impact of certain characteristics of the product or prototype on use actions can become clear. For instance, when a button was not pressed, was it because the action was never intended, or because adjacent text was not understood? Insight into user's perception and cognition can also guide the designers in finding solutions for interaction difficulties. In our experiments, knowing about subjects' perception and cognition was considered important, because it could possibly help in identifying relevant differences between prototypes and the real thing. However, users' perception and cognition are not directly observable. They can sometimes be derived via assumptions of plausibility, or they have to be verbalised by the users. As this verbalising is generally not done spontaneously, subjects have to be asked and encouraged to do so. Even then, only an incomplete and inaccurate account of perception and cognition seems to be supplied.

In the case of the blood pressure monitor concurrent thinking aloud yielded poor verbal reports, for instance because users refrained from verbalising, because they judged their thoughts as irrelevant, or because they couldn't express their activities in words. The general finding that automated behaviour, of which parts of the interaction with the blood pressure monitor consisted, can not be expressed verbally (Ericsson and Simon 1993) was confirmed in this experiment. Subjects often fell silent when getting into difficulties, and this is just when additional information about perception and cognition is most wanted. Retrospective questioning yielded more verbalisations. However, one never knows whether these verbalisations reflect user activities at the time of usage. Sometimes users seemed to reason anew, trying to justify their actions. In a few cases retrospective information conflicted with what actually happened during usage. For example, a subject reasoned why the cuff was applied around the left arm, whereas in fact she had applied the cuff around her right arm. For a more extensive discussion about possibilities of thinking aloud, see Rooden (1998).

These difficulties of eliciting users' perception and cognition limited the possibilities of identifying featural and functional characteristics of the prototypes and the product, by which certain use actions were triggered.

Artifacts in usability research

Apart from difficulties in eliciting users' perception and cognition, some other aspects of the research inhibit a clear view of users' activities in product/prototype usage.

When subjects perform two related tasks, such as operation of a prototype and subsequent usage of the real product, what happens in the first task has an effect on what happens in the second task. This limits the possibilities for intra-individual comparison. Previous experience and familiarity with the product at hand and similar products play an important role in product usage. In the trials, parts of the product, such as the cuff, were recognised and were not inspected in detail. When asked why subjects did what they did, they often said they just knew, or that they remembered it from previous experiences. Often they could not express where these experiences and knowledge came from. Laboratory settings may also inhibit normal usage. Subjects may be nervous, or may try to appear knowledgeable or competent. In our experiment, there was no opportunity to study a manual. Some subjects mentioned that they normally would have read the manual. However, this can also be a way to excuse themselves for 'mistakes' in the interaction. It was deliberately decided not to include a manual and to have subjects focus on the product and its use cues.

It is also worth mentioning that subjects could not cooperate with someone else, which might well be the case at home.

14.6 USERS' TRIALLING WITH MAXIMUM EFFECT

Despite difficulties in researching this topic, some insights were gained on how to carry out users' trials with prototypes. These may be summarised in three categories:

1. Construction of the prototypes

Some suggestions on how to construct and apply prototypes were given in section 14.3, and are repeated here. Although these suggestions are not revolutionary, they may be helpful in making decisions in a design process.

- Attempt to show all sides of the design in a drawings-prototype, for instance by presenting a set of drawings, or a computer animation. However, it should be realised that certain subjects are not familiar with 'reading' drawings.
- Hidden information should be communicated to the subjects.
- Make changes in the appearance which occur as a result of simulated action, visible in the prototype, for instance by preparing some extra drawings.
- Physical aspects can be left out when there is only one obvious way of operation, but this should be done warily, since it is not always clear beforehand.
- Functioning prototypes are preferred over non-functioning ones. When applying non-functioning prototypes it should be realised that users' goals may be different, and

that certain use actions may not be mentioned, or that the order of actions may be inaccurate.

- Add users' trials with existing 'similar' functioning products to the design process, to anticipate interaction difficulties connected with the functioning of the product.
- Simulated feedback should be supplied by an anonymous person, not by the test leader. This makes it easier to stick to the response times of the real product. This may also be facilitated by a computer simulation.

Known shortcomings in the prototypes should be mentioned to the subjects at an appropriate moment during the trial. The aim of users' trialling is not to confuse subjects or to act as an intelligence test. It is the simulation which is being tested, not the subject! This requires an active role of the test leader.

2. Active role of the test leader

A passive role is often advocated for the test leader. This may be true when usage of a real functioning product is to be observed. However, during a design process the aim is to gain information on user activities, and a different role is required of the test leader. During a trial the test leader should intervene, when he or she thinks that by doing so, more information can be gained. This intervention may consist of asking subjects for details when it is not clear what subjects are doing, or helping subjects when they are stuck. Experience in users' trialling will help the test leader to know when to intervene. Some suggestions were given in section 14.4:

- When subjects operate a drawing, and use actions have to verbalised, an active role by the test-leader is desired. He or she should probe for specific information and details during the operation.
- Additional means of expression, such as gesturing and pointing, should be encouraged.
- When subjects wrongly mention that they expect the real product to differ from the prototypes, by adding or changing certain characteristics, they should be corrected.
- Subjects should be told at an appropriate moment what is being simulated, and by what, in the prototype. For instance, when subjects have identified the LCD, they can be told that information on the display will be simulated with cards.

3. Prediction important

It seems unwise to assume that operation of the prototype resembles future usage of the product, and that the design of the product should be tuned to all observed user activities in trials with prototypes. Common sense and 'expert knowledge' may help to predict which of the difficulties happened because of characteristics of the prototype or the set up of the trials. It should also be understood that new difficulties may emerge with the real product, which 'experts' may to some extent be able to predict.

14.7 DIRECTIONS FOR FUTURE RESEARCH

Generalisation of results from users' trialling with prototypes to actual product usage remains an interesting and challenging topic for further research. A series of experiments may build a body of findings, shedding light on the generalisability from the findings with the blood pressure monitor to other products. Studying the capability of experts to handle results from users' trials with prototypes is useful, to find out to what extent they can judge which of the observed difficulties are due to prototyping aspects, and which are due to the design itself, thus requiring design improvements.

14.8 REFERENCES

Ericsson, K.A. and Simon, H.A., 1993. *Protocol Analysis*. MIT press, Cambridge MA.
Kaulio, M., 1997. *Customer-Focused Product Perspective*, PhD thesis. Chalmers University of Technology, Göteborg.
Rooden, M.J., 1998. Thinking about thinking aloud. In *Contemporary Ergonomics*, M. Hanson (Ed.), Taylor & Francis, London.
Ulrich K. and Eppinger S., 1995. *Product Design and Development*. McGraw-Hill Inc, N.Y.

CHAPTER FIFTEEN

Best Practice isn't Always Good Enough

STEPHEN TRATHEN, DON CARSON

Faculty of Environmental Design, University of Canberra, PO Box 1

Belconnen ACT 2616, Australia

and

PETA MILLER

Health Access, PO Box 3021 Weston Creek ACT 2611, Australia

ABSTRACT

This chapter outlines the design of a device to assist in computer aided interviewing for use by employees of the Australian Bureau of Statistics. The product was developed using a design/evaluation iterative loop, with concepts being checked with users throughout the design process. Despite following apparent best practice in terms of tackling usability issues, the product required significant amendments after being tested in the field. The project highlights the need for project team input to the selection of focus groups participants, and the importance of verifying smaller-scale user input with results from larger-scale field user trials. Valuable lessons were learned which served to underline issues raised in Part 2 of this book, regarding the design of usability trials.

15.1 THE CLIENT AND THEIR NEEDS

The Australian Bureau of Statistics (ABS) is a Commonwealth Government entity with statutory responsibilities relating to the collection and analysis of various forms of statistical information. For many types of projects, data has traditionally been collected by interviewers visiting selected households and interviewing respondents using paper questionnaires. More recently, however, the ABS has been involved in the development and trialling of Computer Assisted Interviewing (CAI) for ABS interviewer administered surveys. CAI involves the use of notebook computers to administer the questionnaire during household visits or over the telephone. These personal interviews typically last between 10-20 minutes and can be conducted outside or inside, either seated or standing.

In 1995, the ABS commissioned an investigation of the introduction of CAI. This investigation identified occupational health and safety risks associated with the use of notebook computers in CAI, and recommended further assessment and management of these risks. One of the major findings was that there was an undue injury risk caused by the static load strain associated with supporting the weight of the computer by the interviewer, especially during standing interviews. After considering the recommendations, the ABS initiated a follow-up consultancy to develop practical strategies to minimise the occupational health and safety risks associated with the use of notebook computers in CAI. This consultancy, which was secured by the authors, is the subject of this chapter.

15.2 THE PROJECT

Initial discussions with the ABS indicated that a device was needed to support the weight of a notebook computer when used in a standing position. The consultancy brief was to determine first whether a suitable ergonomic aid was currently available, and secondly, if no such aid existed, to design, develop and produce one.

Various important considerations were communicated by the ABS in the tender documents and in briefing sessions after the consultancy was awarded. Primarily, the device needed to alleviate the identified hazard, that is, the static load strain imposed by holding the notebook computer. Another important factor was that the device should not interfere with the interview process, needing to be lightweight, portable, stable and easy to use. The design would also need to be aesthetically and practically acceptable to the users, the ABS interviewers, and to accord with the professional image of the ABS in the view of interview respondents.

Timing was a major constraint on the consultancy. The ABS required eighty units in the field by early January 1996, allowing only four months to conduct the project. The design team were conscious that, should design of a new device prove necessary, manufacture and delivery timing would span the Australian summer Christmas holiday period.

15.3 OUTLINE OF THE CONSULTANCY PROCESS

Both the design team and the ABS were committed to a participatory model of design that considered the ergonomic requirements of users — their safety, efficiency and comfort. A range of techniques were used, including focus groups, surveys and individual interviews to allow the team to understand users' agendas and concerns, and to benefit from their experience. This process actively fostered participation in, and ownership of, the design solutions. ABS management was supportive of this methodology, and the team worked to ensure this approach permeated the entire project. The key stages of the consultancy can be summarised as follows:

1. preliminary meeting with stakeholders;
2. literature and product search;
3. mock-up production (with input from small focus groups of users);
4. mock-up refinement and prototype development;
5. ergonomic assessment of selected prototype (incorporating a small field test and resultant modifications);
6. production of eighty devices for use in the field; and
7. revision of a device.

15.4 PRELIMINARY MEETING WITH STAKEHOLDERS

We identified the key stakeholders in the project as:

- ABS management;

- ABS staff generally and the user group in particular;
- union representatives;
- potential manufacturers; and
- ourselves, as the consulting team.

The planned change from traditional paper-based data collection to Computer Aided Interviewing (CAI) was a significant deviation from current work practices. This process was complicated by the short project time frames in which the ABS management had to deal with a myriad of associated technical and industrial issues and problems.

Despite these pressures, ABS management recognised the importance of full consultation with all stakeholders throughout the process. The employees' union was represented throughout the project and in later stages played a pivotal role in the process of design development. Union representatives were concerned primarily with the health and safety implications of the CAI process for their members. Staff were generally supportive of the introduction of the new technology and the concept of a support device. However, as the project progressed significant regional variations in levels of exposure to CAI emerged and these differences later proved to be a influencing factor in user-reactions to design options.

While potential manufacturers were not represented in person their likely needs and concerns were considered throughout the process.

Initial meetings were held to gather information about the project and to understand how the device would interact in the day to day work of the interviewers. The team met with the stakeholders to discuss current relevant work practices, interview techniques and explore perceived user requirements. The team also met with senior management to familiarise themselves with the major economic and structural issues and constraints associated with the project.

15.5 LITERATURE AND PRODUCT SEARCH

The design team conducted a literature search to assess current research on ergonomic issues and considerations associated with the use of lap-top computers, but found little of immediate relevance.

The team also undertook a product review to establish if any device had already been developed which could be adapted to meet the users' requirements. While the assessment process revealed that there were no currently available products which adequately met the design requirements, the team was able to use ideas and adapt mechanisms for inclusion in the new design. These included:

- photographic equipment (including tripods, studio light stands and slide projector stands);
- musical equipment (including sheet music stands, keyboard and percussion stands); and
- rehabilitation products (including walking sticks, with mono-, tri-, and quad-feet).

15.6 MOCK-UP PRODUCTION

After investigating the range of products identified in the product review, the team constructed three simple, non-adjustable mock-ups which were all fairly similar in design. The mock-ups represented a range of footprints, comprising a mono-pod, tripod, and quad-stick.

During the next phase of the project, focus groups were convened to obtain initial user criteria for the device. Participants in the groups were selected by ABS management with the aim of representing those with an interest in the product, namely, the interviewers, management and union. Focus group members had experience in traditional work methods, but none had yet used CAI methods.

At each focus group, the team outlined the potential health and safety problems associated with CAI use; the project objectives, the design team's role, and the methodology. The team emphasised the critical need for ongoing user input and feedback to ensure that the project objectives were met. A combination of the use of scenarios, two dimensional concept sketches and mock-ups was used to help prompt discussion and to encourage exploration of potential problems and their solutions.

All focus group members were given equal status and opportunity to contribute and their comments were recorded without judgement or discussion until the end of the session. A diverse range of user concerns and concepts was generated from this process, which was later transcribed by the team and used to develop the preliminary design criteria. These were that the device should be:

- of a simple design, requiring minimal assembly time. Simplicity was valued more highly than the compactness which could be gained from including more telescoping and other joints;
- portable, perhaps incorporating a shoulder strap for ease of carrying;
- in keeping with the interviewers' professional image;
- adjustable but able to be fixed in a position best suited to the individual operator;
- able to stand unaided on a regular surface;
- light weight; and
- regular in shape, with a minimum number of awkward protuberances.

Participants also emphasised that any necessary attachments to the lap-top would need to be flush with the moulding in order to protect surfaces, such as the dining room table tops which often became their work space in interviewees' homes. The groups did not give priority to the ability of the device to be used on uneven surfaces or rough terrain, or the ease of use when sitting at the device, as they believed interviewers would modify their routine to suit different circumstances.

In addition to focus group input, the preliminary design criteria were developed with reference to the anthropometric databases for reach, posture, and strength. These were then used to establish optimal weight and adjustment mechanisms for the device.

Prototype refinement and prototype development

Using these preliminary design criteria, two prototypes were developed to present to a group of ABS interviewers who had participated in the CAI field trials. This meeting was crucial in the overall design process, as these were the first focus group members who had experience with CAI. This group articulated problems they had encountered in using the lap-top computer to date and, before being presented with the prototypes, were led through a process of developing their potential design criteria. Interestingly, the criteria obtained from this process closely aligned with the criteria developed by the earlier focus groups, and the response to the prototypes was very positive.

Ergonomic assessment of selected prototype

The prototypes were assessed against the refined user design criteria and, as the Rapid Upper Limb Assessment (McAtamney and Corlett 1993) had been used to establish the initial risk of undertaking CAI without a support device, this method of assessment was also used in ergonomic evaluations of the prototypes.

Two identical prototypes were then field tested by a small number of users. Feedback from the small field trial agreed with that obtained from the earlier focus groups, with interviewers again placing highest priority on simplicity.

Initial ergonomic design criteria had been to develop the device to be able to adjust in height to suit from the 5% of female users to the 95% of male users. During the ongoing liaison and consultation process, the ABS indicated that user population was predominantly female, aged between 45-54 years of age. A total of 60% of the entire user population was aged above 45 years of age, with 75% women over the age of 35. This influenced the anthropometric data required, particularly in determining the height range of the device.

The design team, in conjunction with ABS management and user representatives, decided to reduce the height adjustment range, in order to decrease the overall weight and size of the device. The new height adjustment range was therefore modified to suit the 5% of female users to the 90% of male users, with the client's acknowledgement that users outside that anthropometric range would require customised devices.

Production of 80 devices for use in the field

Eighty of the devices were produced by the required deadline and equally distributed between the states of Victoria and Queensland for use in the field during a national survey. The eighty interviewers were then asked to complete a user feedback survey (this sample represented 10% of the total number of future interviewers). Interviewers were asked to rank the relative importance of attributes of an 'ideal' device, to rate the features of the actual prototype device and to comment on aspects of the device and suggest improvements.

The survey results differed from earlier feedback provided by the focus groups. Both the earlier focus groups and the smaller field trial identified ease of use and minimum adjustments as important design features. The device produced for field use reflected these criteria and was designed with few folding or retractable parts. However, these features also made the device somewhat bulky and consequently, conspicuous (see Figures 15.1 and 15.2).

Figures 15.1, 15.2 and 15.3 Device for field use.

Whereas earlier groups had accepted this, interviewers in the larger field test found the bulk a problem. While ranking other attributes highly, 68% thought the device did not present an appropriate professional image. This appeared to be related to the visibility of the device. For instance, many interviewees reacted negatively to the interviewers as they thought the device was for sale and the interviewers were salespeople. This initial resistance was not conducive to co-operation with the interviewers, and decreased the chance of the interviewer being invited indoors to conduct the interview.

The field-trial also revealed gender-based differences in acceptance and use of the device. Men used the device more frequently than women, with 46% of men reporting use of the device compared with only 29% of women. When asked to rank attributes of the device, in terms of importance, other gender differences emerged. Males ranked (from most to least important) attributes as follows: stability, ease of adjustment (equal ranking), portability, weight, size and storage, whereas females ranked portability as most important, followed by weight, size, ease of adjustment, storage and stability (equal ranking).

Revision of device

The feedback from the larger survey lead to a revised set of criteria for the device. As the device's visibility was problematic, it was decided to incorporate the revised device within a carry bag that would also hold the lap-top and material required by the interviewer. This amalgamation of the device with the more familiar visual appearance of a briefcase or business case addressed the negative interviewee perceptions associated with the earlier design (see Figure 15.3).

15.7 ANALYSIS OF THE PROJECT

Despite the design team's commitment to a participative process of design, significant differences emerged between earlier interviewer input and the outcomes of the major field test.

In retrospect, this problem may be traced to the structure of the earlier focus groups. The composition of these groups proved unrepresentative of the user population. Membership was determined by ABS management, rather than by the design team, and was influenced by factors other than those the team would have used. For example, the union required a certain number of union representatives to be included, and the groups did not represent the actual gender ratio of the interviewers. Focus group membership was usually split equally between males and females, whereas approximately 80% of interviewers were women.

The problem underlying these discrepancies lies in the fact that each set of stakeholders had their own perspective on the purpose of the focus groups. Management saw focus groups as primarily a mechanism to involve staff and unions in the process and thereby gain their support. Unions saw the groups as another forum to air their concerns and defend their interests. From a theoretical design point of view, the sole consideration in determining composition of such groups should be the ability of members to contribute to the design process, and specifically to the design of a product which will reflect the needs of the entire user population. In the real world, as in this case study, designers are then faced with the challenge of balancing these competing demands: client expectations on the one hand, and design requirements on the other.

Two main options present themselves in responding to this challenge. First, designers could attempt to keep the focus groups purely centred on design outcomes, and ensure representation on the groups is in keeping with this outcome. Pursuing this option would necessitate finding alternative methods to meet client-specific expectations. In the case study, for example, perhaps union and management objectives could have been met by establishing a steering committee, including representatives from both parties. The committee could then have acted as the forum for issues beyond the scope of the design brief (such as union consultation on the overall change process), thus allowing the focus groups to be governed by design considerations.

Alternatively, designers could accept that non-design based objectives will only be separated from the design process with difficulty. In adopting this approach, designers would then need to compensate for the limitations presented by the situation. For example, in the case study discussed in this chapter, perhaps the number of union representatives could have been balanced by including a similar number of other participants more representative of the overall user population (for example, more women could have been included).

Aside from the implications of discrepancies between designers and clients' objectives in the use of focus groups, stakeholder representation is also an issue. While management, union and interviewers were closely involved, members of the public, the interviewees, were not. Although not 'users' of the design in a narrow sense, the views and reactions of this group proved crucial in the users' perceptions of the device in the field trial.

Obtaining input from the community at large is problematic, especially in situations like this case study, where the target group usually deliberately randomised. While in this case, focus group representation was probably unrealistic, it may have been advantageous to develop an extra step in the process. For example, instead of moving direct from focus groups to a large field trial, perhaps a process of role plays incorporating members of the public could have been included as a possible method of predicting public reaction to the design.

Setting aside this discussion of potential improvements to the structure and process of focus groups, a more fundamental question should be posed. In assessment of the design team, the disparity of outcomes between the design teams initial and later processes cannot be accounted for on the basis of focus group structure alone. The actual applicability of stated views to final design should be questioned, as users' theoretical assessment of their design values and perceptions appear to differ markedly from their real-life reactions in the field.

In fact, this assessment aligns with many everyday experiences — what we say we want, or will do, is not necessarily what we want or do in practice, as any market researcher or pollster can attest. The design team believe this phenomena is especially the case in situations involving client contact, where users' perceptions will be influenced by the those of their clients. This is not surprising given the complexity of the issues involved. The design outcome was influenced by many factors, but the factor which lead to a negative assessment in the field trial was related to the interviewees' emotional responses to the device: an area of design and ergonomics that is currently undergoing a growth in interest and research as indicated by material published by authors such as Crozier (1994) and Jordan (1998). It was important to the ABS interviewers that they put the interviewees at ease, as they realised that first impressions would have a significant impact on whether an interviewer was well received.

One approach to this problem is to reassess the definition of 'users' of the design. The ABS interviewers are matched by those they interview, a receptionist in a waiting area is matched by the clients who approach the desk, or a doctor examining a patient with an ultra-sound equipment. User-centred design must perhaps define users more broadly in order to canvass the entire range of responses and views.

However, the point remains that the limitations of small-scale focus groups to the design process must be recognised. Field trials will always be necessary to reality test their outcomes, and perhaps the moral is, we should not be too surprised if the results do not align with more theoretical frames of input.

15.8 REFERENCES

Crozier, R., 1994. *Manufactured Pleasures: Psychological Responses to Design.* Manchester.

Jordan, P.W., 1998. Human factors for pleasure in product use. *Applied Ergonomics,* **29**, pp. 25-33.

McAtamney, L. and Corlett, N.E., 1993. RULA: a survey method for the investigation of work-related upper limb disorders. *Applied Ergonomics,* **24**, 2, pp. 91-99.

CHAPTER SIXTEEN

Inclusive Design — Design for All

ROGER COLEMAN

Royal College of Art, Kensington Gore, London SW7 2EU, UK

16.1 INTRODUCTION

In this short chapter I shall try to give an overview of the complex subject of population ageing, something which could eventually effect every human population. Against this backdrop I will introduce a range of market and design strategies that I believe to be appropriate, and go on to offer some short case studies, of both professional and student work, as an indication of present and future trends in design.

In terms of technology, if we try to look into the future we quickly enter the realm of science fiction. However, from a demographic standpoint we have much harder data to rely on and can be quite confident in our predictions about what the future shape of our societies will be. Demographic patterns are persistent over time and change slowly as factors work their way through the population, and as far as the developed world is concerned one of the strongest trends is towards older societies. The reasons are complex, and the social implications both interesting and far reaching. In essence we are living through a dramatic change in the age structure of populations around the world; in some countries the process has barely begun, in others it is close to completion.

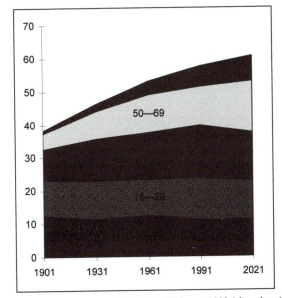

Figure 16.1 The UK population by age, 1901–2021. (Coleman 1993a) based on (Dennis 1991).

In North America the average couple has more parents than children; by 2020, every second European adult will be over 50; there is already incontrovertible evidence that youth markets are shrinking in absolute numbers and in value while older people command an increasing proportion of wealth and disposable income — currently some 65% of UK assets and 35% of consumer spending is controlled by people aged 50 and over. The UK was the first country to experience an industrial revolution and one of the first to age.

One hundred years ago the over 50s were a small proportion of the population, but by 2020 their numbers will have seen a four-fold increase, while the number of young people will have remained static or even fallen. This is a radical, unprecedented and probably irreversible change which Peter Laslett has called the 'Secular Change in Ageing' (1996). For the UK, it has been possible to reconstruct trends in ageing and longevity from around 1,000 AD onwards, revealing how an earlier pattern of short life-expectancy and low numbers of old people will have been reversed over a period of about 200 years. In the UK this process is almost complete and as a result, the age balance of the population has changed out of all recognition.

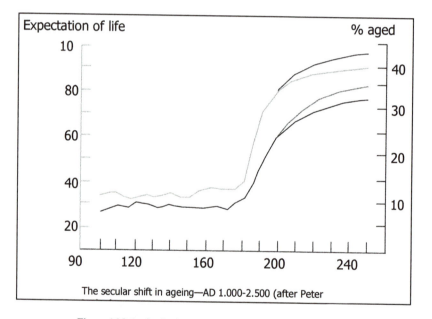

Figure 16.2 Ageing in the UK, AD 1,000–2,500 (after Laslett 1996).

At the end of the 19th Century, male life expectancy in the UK was less than 50 years, the majority of the population was aged under 30, and the proportion surviving to 70 tiny; 100 years later, most people live to enjoy their 70th birthday and can reasonably anticipate a long period of life after work. The same pattern is repeated around Europe. From Sweden in the north, Europe's most mature population, to Spain in the south, we see the same tendency, although with considerable variation due to cultural, economic and other differences. This is accompanied by a strong growth in the proportion of people aged over 60 and, importantly, those over 80, who are likely to be significant consumers of health and welfare services.

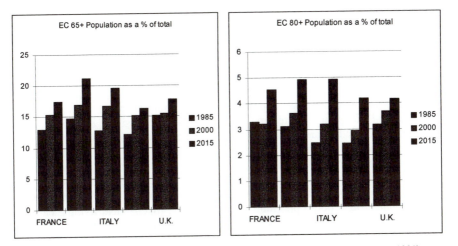

Figure 16.3 and 16.4 The 65+ and 80+ population in Europe, 1985–2015 (after Yoxen 1991).

What is happening in Europe is indicative of a radical change in the structure of populations around the world, a change that has passed almost unnoticed until recently, but will be looked on as one of the most dramatic ever to affect our societies. Economic development is closely associated with this change which is characterised in industrialised societies by a steady decrease in birthrates, which in turn can be linked to changing expectations among women, the transformation of traditional cultural values, increasingly individualistic attitudes and behaviour, and a complex mix of other factors.

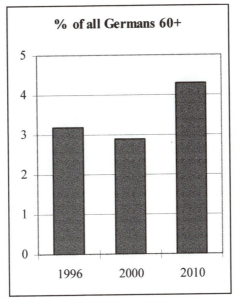

Figure 16.5 The 60+ population in Germany, 1996–2010
(after Naegele 1997) based on German Office of Statistics 1994.

In the West, growing evidence suggests that the effect of improved diet, medicine and living conditions is to prolong not just life expectancy but active life. If the majority of older people can remain economically independent, spending their money on an improved quality of life and thereby creating demand for new goods and services, then what is too often perceived as the old people 'problem' could turn out to be an opportunity (Henley Centre 1987; Buck 1990).

However, it is wrong to assume that this will happen universally. Along with this optimistic trend in the industrialised countries, there is also a growth in the number of socially disadvantaged older people, predominantly women and members of the lower social classes, suggesting that population ageing could increase disparities in living conditions and life quality. Increasingly older people live in single households. Already, the traditional extended family offers little support to such people in their old age, and the alternative, institutional care, is not welcomed. Instead there is a strong and growing desire to live independently for as long as possible, which remains even when people are in poor health, or in need of constant care (Naegele 1997). Making it possible for people from all levels of society to achieve the goal of independent living presents us with as many challenges as it does opportunities. Importantly, independent living means living a 'normal' life; people do not want special products or to be separated out from the rest of the community, and this is something that manufacturers and designers must take heed of now and in the future (Smith 1990).

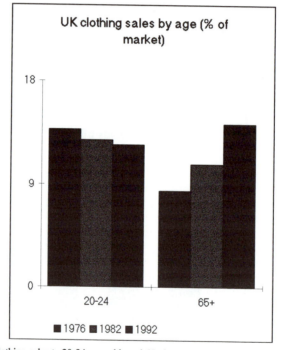

Figure 16.6 UK clothing sales to 20-24 year olds and 65+ by percentage of total sales, 1976-1992 (from survey figures supplied by TMS Partnership, London, August 1993 — base sizes 1976-24, 597; 1982-31, 889; 1992-51, 882).

The overall result of these far reaching social developments is a new matrix of wants, needs and aspirations that is beginning to determine older people's behaviour. Increasingly they are looking on later life as an opportunity for self-realisation, a time to do many of the things that were not possible while they had responsibilities in the shape of young families and careers. In order to achieve that goal certain things become important: managing time to make the most of these new possibilities; managing the ageing process itself, by staying fit and healthy and later by finding ways to remain independent for as long as possible; living with the new family structures that are the consequence of lower birth-rates, increased divorce and the growth of single parent

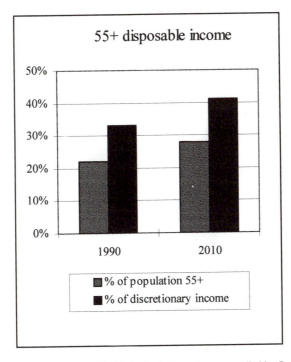

Figure 16.7 55+ disposable income in The Netherlands (from figures supplied by Ger Tielen of The Netherlands Platform for Older People and Europe).

families; understanding how to manage money and ensure financial security over many years; and maintaining the social networks of family and friends that are so important to us all (Henley Centre 1996a, 1996c).

People in their 50s and 60s are experienced and discerning consumers with rising expectations of products and services, which means that attitudes will have to change radically among all those who design, manufacture and supply them. Paradoxically, older consumers barely feature in market research, are generally ignored by advertisers and are currently shunned by manufacturers and retailers alike. In addition, there is a persistent and erroneous association of age with disability. The constant references to 'The Disabled and Elderly', that are part of official-speak in academic, government and charitable circles, are as offensive, as they are ill-informed, and only serve to set both older people and people with disabilities apart from the main body of humanity. The reality is very different: the majority of the 8-10% of the population officially recognised as disabled is old, not young, but only a small minority of older people are disabled and fewer still in need of institutional care. In other words, the wheelchair user — the common symbol for 'disabled' facilities — is as atypical of people with disabilities, as disability is atypical of older people.

Conventional assumptions about ability and disability are out of phase with reality in more ways than one. First we tend to think of people as disabled by their physical or mental state, which means that in a sense they carry their problems around with them rather than discover problems when they confront obstacles. Not only does this ignore the fact that our capabilities are constantly changing under the impact of disease, accident, tiredness, ageing, pregnancy, and are affected by what we are doing — pushing a wheelchair or pram, carrying a child — it fails to understand that disability is the result of a mismatch between people and their surroundings, which is all too often the consequence of poor and unthinking design. A mismatch which good, thoughtful design can overcome.

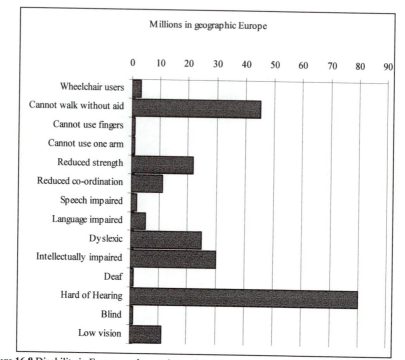

Figure 16.8 Disability in Europe — the numbers of disabled potentially effected by the design of ATMs (after Gill 1997).

The overwhelming majority of people with disabilities are older people with minor impairments which hinder everyday activities. For every profoundly deaf person, for example, there are 80 who are hard of hearing and mostly aged over 60. Similarly, for every wheelchair user there are 15 others who cannot walk far without aid or assistance. There is, therefore, a real possibility of improving the environment to the benefit of large numbers of people, and considerable commercial and social advantages could follow. By removing steps we benefit not only those in wheelchairs, we include the many more people who do not need a wheelchair, but walk with some difficulty. The same is true in relation to hearing, by addressing the needs of deaf people we are likely to go a long way towards benefiting the 10% of the population who are hard of hearing, many of whom miss out on important announcements and safety information.

16.2 DESIGN FOR ALL

What is clear is that with good design we can address the needs of very large numbers of people who are at present disabled in one situation or another by poor or inconsiderate design. The extent to which we can do this will be extended by technology, and in the future we could well reach the point where there is very little need for specialised products. However, if the way in which new technologies are introduced does not take these factors into account we run the risk of disabling much greater numbers of people in the future. We will never finally eradicate old age or mental and physical impairments, on the contrary what the demographics tell us is that we will <u>all</u> have to live with them to some extent, and should embrace such facts of life as part of the human condition. The important thing is to constantly strive to minimise the impact they have on our lives by designing a world which works for everyone, regardless of ability, and where that is not possible, to use our creativity and ingenuity to design the special products that offer people as much dignity and enjoyment as possible. The answer lies in what is increasingly referred to as 'Universal Design' or 'Design for All', not compensatory gadgets to make up for poor design.

To make this a reality, we have to move beyond ergonomics, beyond ideas of compensatory products and the rather medicalised view that we have of ageing as a disabling process of decline and debilitation. Instead we need new models to make it possible for us to embrace rather than reject or fear the future. With models based on capability, on activity and autonomy, we can begin to measure not the deficits of ageing, but the abilities and activities that can be sustained into old age, and instead of thinking in terms of ageing at a personal and social level as a 'problem', turn these ideas on their heads and recognise it as a gift and an 'opportunity'. In doing so it is important to recognise that ageing is as much a social construct as a physiological fact. Understanding older people now and in the future requires that we see ageing as a dynamic and developmental process, and from this perspective it becomes possible to see older people in a new light (Graafmans *et al.* 1998).

First of all, lifestyle is far more significant than chronological age — a 50 year old and a 70 or 80 year old can easily share the same interests or activities, but have very different capabilities — which means that unless the products and services on offer work well for and appeal across a wider range of age and capability, they will not fully capture the market that already exists and are unlikely to stimulate future growth. Second, older people are experienced consumers who like to shop around, and have more time than the working population for that activity. They love being busy, avoid isolation and loneliness, and assess themselves in terms of what they can do, not what they cannot do. What is important to them is that products and services assist them in maintaining their vitality and interest in life. Third, in a sense older people are moving beyond consumerism. Younger people are more likely to want the latest or most fashionable products, to consume 'things', whereas older people are beginning to realise that — to paraphrase Mae West — its not the years in your life but the life in your years that really matters, and so are looking to fill their time in meaningful and sociable ways. This presents a further challenge to design and manufacturing: delivering goods and services that go beyond functionality to offer pleasure in use and real improvements in life quality for people of all ages (Coleman 1997a; Henley Centre 1996a-c).

Three broad areas of opportunity emerge from this approach. First, the mainstream provision of goods and services in both public and private sectors, and importantly in the leisure and retail sectors, along with the built environments that surround them, can benefit from becoming age-friendly. Here the concept of 'barrier-free' architecture has a role to play in removing obstacles and encouraging the participation of people of all ages and abilities, as does 'universal' or considerate design in everyday objects: packaging that can be opened without taking a knife to it, chairs that are easy to get in and out of, clothing that is light, warm, easy to put on and to clean, but still elegant and good looking; signage and information graphics that are clear and easy to read, controls that are simple and easy to use. These things will broaden the appeal of goods and make life better for all who use them, not just older people.

Second, niche products and services targeted at older people in ways that are positive and non-patronising, like 50+ insurance policies, activity holidays, well designed, adaptable housing and 'Senior Net' products. There are other opportunities waiting to be developed. For instance, the bulk of the DIY and interior decor market is targeted at younger householders, while there is a need for older people to make changes to their homes that will help them live independently. Unfortunately, when thinking about older people we tend to use terms like 'housing adaptation' rather than 'home improvement', giving the subject an unnecessary negative feel. There are already some excellent household products on the market and plenty of scope for developing this sector on the basis of lifestyle and independence. Unfortunately, people are not aware of what products are available, how they could benefit from them, or just how much they could improve their homes with good, age-friendly interior design. The newspapers and magazines that help people do this will be richly rewarded in advertising income, as will be the manufacturers and retailers that older people come to think of as partners in making their lives richer and more independent.

Third, there is a growing market for specialist products and services — from IT interfaces and other assistive products, through home-based services and on to institutional care — that address the needs of the minority of older people with disabilities, and those who require regular care. At present, this market sector is poorly serviced. Health care and welfare services have been reluctant to spend money on research and development, and as a consequence most technical aids have been designed with functionality and economy uppermost and are depressing to look at and stigmatising to use. This is set to change, and manufacturers that seize the opportunity to grow alongside expanding European markets for high quality, well designed assistive products and services will do very well in the future. In particular, those that invest in good design and seek to break the prevailing institutional mould will find their products and services have a broader appeal and marketability outside the institutional sector.

A number of different design strategies and approaches have been developed in response to these facts and a few examples are given below.

16.3 UNIVERSAL DESIGN

In the USA, the concept of universal design has been developed over a period of 20 or more years, and has achieved considerable international recognition. In the summer of 1998 the Japanese universal design Consortium published the first issue of a new journal devoted to the subject, and an international conference on universal design attended by over 400 delegates was held at Hofstra University, New York and organised by the Centre for Adaptive Environments, Boston. The architect and designer Ron Mace devoted much of his working life to developing and promoting the concept of universal design, and a growing number of institutions and design courses have embraced this concept, exerting a significant influence on the design profession and on government and industry. The Americans with Disabilities Act (ADA) reflects this, making access to environments and services a right for all American citizens, while companies are beginning to realise that universal design can improve the quality of their products for all users. Universal design is best described as a methodology that maximises the number of people who can use a building, a product or other object, and attempts to reduce the need for separate accessible or adaptive design solutions. It does this by going beyond the needs and abilities of average, healthy adults to include those with motor and sensory disabilities, children and older adults during the design process (Trachtman and Reagan 1999).

16.4 THROUGH OTHER EYES

A leading proponent of universal design is Patricia Moore, whose work builds on remarkable experiment carried out in her 20s when, based at Raymond Loewy's NY offices, she spent three years (1979-82) travelling throughout US and Canada visiting over 200 cities and experiencing life disguised as an 85 year old woman. The discoveries that she made shaped her thinking about design and were influential on the growing universal design movement in America (Moore 1985).

16.5 TRANSGENERATIONAL DESIGN

Professor James Pirkl and colleagues at the University of Syracuse developed the concept of Transgenerational Design — to describe products, services and environments that meet the needs of people from a wide range of age groups and with differing needs and abilities. They have produced a series of guidelines and strategies for applying this concept in the fields of design, journalism, advertising, retailing, employment policy and marketing (Pirkl 1993; Pirkl and Babic 1988).

16.6 REHABILITATION DESIGN

David Guynes of Guynes Design has been working in partnership with Patricia Moore on innovative approaches to the design of health-care and rehabilitation facilities which focus on encouraging independence and an early return to the community as opposed to long-term institutional care. The basis for their work started about 12 years ago with the development of Easy Street™, a microcosm of the local community within a health-care setting, which is used as a training area for patients to confront the obstacles, both physical and cognitive, of their local community (Coleman 1997a).

16.7 HEALTHY INDUSTRIAL DESIGN

HID™ is a concept developed in the UK by Alan Tye RDI. In response to growing awareness of some of the problems associated with the use of products and environments e.g. repetitive strain injury, he has developed a strategy which draws on insights gained from a range of body-mind techniques and applied these to the design of products and environments, with the intention of enhancing the health and enjoyment of the user through appropriate movement (Coleman 1997a).

16.8 DESIGN BY STORY-TELLING

The international design company IDEO has worked on a number of products which address older markets. The company is known for its innovative approach both to design and to the use of the latest technology in achieving user-friendly solutions to practical problems. A key to its success lies in the effective use of a five-step design process based on scenario building. By setting themselves to first understand and observe the potential user, and then visualise scenarios around a range of users they can identify new product opportunities and ensure that the widest possible range of users are considered in the final design (Coleman 1997a; Moggridge 1993).

16.9 DESIGN FOR A BROADER AVERAGE

Maria Benktzon and Sven-Erich Juhlins of the Swedish consultancy Ergonomi Design Gruppen have demonstrated that by applying the concept of 'the pyramid of needs' it is possible to design for a broader range of needs, including the special needs of specific groups within the population, resulting in products which offer a combination of performance, functionality and aesthetic appeal. These methods have been successfully applied to the design of hand tools, workplaces, catering equipment for airlines, walking sticks and ambulance stretchers (Coleman 1997a; Benkzton 1993).

16.10 THE OWL MARK™

The Centre for Applied Gerontology at the University of Birmingham in the UK, works to impart a better understanding to designers and manufacturers of the needs of older people. Recognising that accreditation of products for older people is important it has developed its own quality mark, the Owl Mark, and supports its investigative work with a panel of older consumers the 'Thousand Elders' (Nayak 1998; Coleman 1998).

This list is by no means exhaustive, and the most interesting thing to note is the growing convergence on the concepts of 'Design for All' and 'Barrier-Free Design', not as prescriptive solutions — a one-size-fits-all will never be successful in the market place and will not solve all our problems — but as important aspirations and key aspects of good design in general. In confronting the implications of population ageing, design for and in the future will have to give high priority to meeting the needs and aspirations of users of all ages to create an integrated and inclusive society in which old and young can enjoy the highest levels of independence and life quality.

16.11 CASE STUDIES

Universal design

Good Grip kitchen tools are designed to be comfortable and easy to use and hold. The basic range includes most standard items and is being constantly extended to cover all cooking activities. A garden range is also being developed, beginning with fork, trowel and other small items. A key feature of the product is the large diameter handles, which are made from non-slip Santoprene rubber. These are easy to hold, soft to handle, and give better leverage in use. The design of the handle distributes the gripping force and by doing so minimises the amount of hand strength normally required and offers a comfortable, cushioned grip for everyone, regardless of age or ability.

Good Grips were designed by Smart Design of New York, and produced for Sam Farber by the OXO company, part of the U.S. General Housewares Corporation. The concept was inspired by the problems Sam's wife Betty experienced with her arthritis but the universality of the design means that top professional cooks use Good Grips, upmarket kitchen shops and major supermarket chains stock the range, and the volume of sales ensures that all the items are modestly priced.

Breaking the institutional mould

Bisterfeld und Weiss is a German manufacturer specialised in mainstream furniture products for the contract market. The company identified a market opportunity for furniture for the home and for more institutional settings. Their first response was an easy chair, with a footrest, arms that fold down, and a side pocket for personal items. That product, though well received was, they felt, still institutional in appearance, and had some ergonomic shortcomings. The designer, Arno Votteler, wanted to go one step further and produce a range of furniture which would make it possible to create an attractive personal environment, especially for people who live predominantly in one room. The range was designed to bring functionality to the domestic environment and the warmth and taste of the home into institutional settings and sheltered housing. Design and development was carried out over a two-year period. All items were given practical tests in hospitals and other settings, and user-tested for functionality and acceptability. Styling and detailing were co-ordinated to produce a coherent range, and form, colour and materials were carefully harmonised to communicate safety, stability and comfort, without compromising the objective of breaking the institutional mould.

Age-friendly packaging with universal appeal

Work by Gavin Pryke (Royal College of Art Ceramics & Glass) and Bente Syversen (Ravensbourne College of Design & Communication), both winners of the 'New Design for Old' section of the RSA Student Design Awards, goes a step beyond purely ergonomic considerations to offer a better quality product with universal appeal. Starting with the specific needs of older people, both ergonomic and lifestyle based, these two students rethought glass and plastic packaging in ways which address both the gripping, twisting and opening problems that older people encounter with much conventional packaging, and combined these insights with other aspects to produce on the one hand a better-looking glass jar, and on the other a combined pack/dispenser for instant coffee.

New vehicle typologies

At the Royal College of Art in London, vehicle design students have been developing concepts for new types of vehicles, particularly for use in our crowded cities. The combination of growing numbers of older drivers, single households and environmental concerns to reduce the number of vehicles on the roads and the pollution they produce presents us with a very knotty problem. Increasingly, older people rely on their cars to help maintain independent lifestyles, unless they are offered alternatives to car ownership they will be reluctant to stop driving and if they are forced to governments will have to meet increased welfare and care bills. The alternative is new types of vehicles — low-impact, small affordable taxis; shuttle buses; easy access trams; even driverless mini-buses and cabs linking home, hospital, public services and so on — all running on electricity or other low-pollution fuels. In the longer term future vehicle manufacturers will have to think in terms of offering mobility for life rather than selling individual vehicles. However if that happens, one of the driving forces, if you will excuse the pun, will be the growing number of older people who aspire to live independently, in their now homes for as long as possible (Sixsmith and Sixsmith 1993; Coleman and Harrow 1998).

16.12 REFERENCES

Buck, S. (Ed.), 1990. *The 55-Plus Market: Exploring a Golden Business Opportunity*. McGraw-Hill, London.

Benkzton, M., 1993. Designing for our future selves: the swedish experience. *Applied Ergonomics*, **24**, 1.

Coleman, R., 1993a. A demographic overview of the ageing of first world populations. *Applied Ergonomics*, **24**, 1.

Coleman, R., 1993b. *Designing for Our Future Selves*, R. Coleman (Ed.), RCA.

Coleman, R., 1997a. Breaking the age barrier. *RSA Journal*, Nov/Dec (1).

Coleman, R., 1997b. *Design für die Zukunft*, R. Coleman (Ed.), DuMont Buchverlag, Cologne.

Coleman, R., 1998. *The Bernard Isaacs Memorial Lecture Journal of the Israel Gerontology Society*. For Eng. Language text see DesignAge website at http://DesignAge.rca.ac.uk/.

Coleman, R. and Harrow, D., 1998. *A car for All or Mobility for All?* Unpublished paper given to ImechE conference, A Car For All, London, see http://designage.rca.ac.uk/da/texts/Carforall/carforall1.html.

Dennis, G., 1991. *Annual Abstract of Statistics*, G. Dennis (Ed.), HMSO, London.

Gill, J., 1997. *Access Prohibited*. RNIB, London.

Graafmans, J., Taipale, V. and Charness, N., 1998. *Gerontechnology: A Sustainable Investment in the Future*. IOS Press, Amsterdam.

Henley Centre, The, 1987. *The New Old*.

Henley Centre, The, 1995. *Family Expenditure Survey*, 1995-6.

Henley Centre, The, 1996a. *Leisure Tracking Survey*.

Henley Centre, The, 1996b. *Consumer Markets 2000*.

Henley Centre, The, 1996c. *Planing for Social Change*, 1996-7.

Laslett, P., 1996. *A Fresh Map of Life: The Emergence of the Third Age*. 2nd edition, Weidenfeld and Nicolson.

Moggridge, W., 1993. Design by story-telling. *Applied Ergonomics*, **24**, 1.

Moore, P., 1985. *Disguised*. World Books, Waco, Texas.

Naegele, G., 1997. Deutschland wird älter. In *Design für die Zukunft*, DuMont Buchverlag, Cologne.

Nayak, U.S.L., 1998. Design participation by the Thousand Elders. In *Gerontechnology: A Sustainable Investment in the Future*, IOS Press, Amsterdam.

Pirkl, J.J., 1993. *Transgenerational Design: Products for an Aging Population*, Van Nostrand Reinhold, N.Y.

Pirkl, J.J. and Babic, A., 1998. *Guidelines and Strategies for Designing Transgenerational Products: An Instructors Manual*. Copley Publishing Group.

Sixsmith, A. and Sixsmith, J., 1993. Older people, driving and new technology. *Applied Ergonomics*, **24**, 1.

Smith, D.B.D., 1990. Human factors and aging: an overview of research needs and application opportunities. *Human Factors*, **32**, 5, pp. 509-527.

Trachtman, L. and Reagan, J., 1999. Designing for the 21st century. In *Proceedings of a Conference*, in preparation for publication in 1999. See http://www.adaptenv.org/21century/ for online proceedings.

Yoxen, E., 1991. *Demography Factbook Centre For Exploitation of Science and Technology*, E. Yoxen (Ed.), London.

DesignAge maintains a special collection of c.1300 items on design and ageing — the Helen Hamlyn Database, abstracts of which can be consulted via the Internet at http://DesignAge.rca.

CHAPTER SEVENTEEN

Inclusive Design

PATRICK W. JORDAN

Philips Corporate Design, Building W, Damsterdiep 267,

P.O. Box 225, 9700 AE Groningen, The Netherlands

17.1 INTRODUCTION

The professional design and human factors community has begun to develop an awareness of issues connected with the design of products for disabled people. This is, perhaps, rooted in a sense of the moral imperative that lies behind designing for the disabled. After all, the social aspect of designing for the disabled is grounded in "... the democratic values of non-discrimination, equal opportunity and personal empowerment" (Mullick and Steinfeld 1997).

Manufacturers are also beginning to see the commercial benefits associated with designing for the disabled. Coleman in the previous chapter notes that the numbers of elderly people in the Western world have increased dramatically over recent years and are likely to go on increasing for the foreseeable future. These people tend to have above average disposable incomes, and are thus an important target group for manufacturers. They are also a target group which contains a higher than average number of disabled people. Thus, designing for disability is central to meeting the needs of this group.

However, disability is not merely confined to the elderly population. There are many younger people who are disabled — perhaps from birth or as a result of a subsequent illness or accident. Unlike the elderly, these people may not have high disposable incomes. In fact, it is more likely that they will have lower than average spending power. Because of this manufacturers may view the younger disabled as an unattractive target group — a marginal group which is difficult to cater for and which will provide little commercial benefit for the companies catering for them.

In this chapter, it is argued that such a view is flawed and is based on a lack of understanding of people as a whole and disabled people in particular. An approach — known as 'Inclusive Design' — is advocated. This approach aims to take a holistic view of users. It recognises that people differ from each other in many different ways and sees any particular disability simply as one of the dimensions along which people can differ. It also recognises that designs which suit people with particular disabilities may also bring advantages, or at least not create disadvantages, for the able bodied. This means that design for the disabled need not always be seen as an alternative to designing for the able-bodied. Rather, in commercial terms, the production of products that are suitable for both able-bodied and disabled is simply a way of extending the potential customer base for a product.

17.2 PRODUCT BENEFITS AND PEOPLE CHARACTERISTICS

It has been proposed that, broadly, products can bring four different types of benefits or 'pleasures' to their users (Jordan, Chapter 21):

- Physio-pleasure — pleasures to do with the body and the senses.
- Socio-pleasure — pleasures to do with inter-personal and social relationships.
- Psycho-pleasure — pleasures to do with the mind.
- Ideo-pleasure — pleasures to do with values.

In order to create products that deliver such benefits, those involved in the product creation process must have an understanding of the characteristics of their potential customers. This understanding should be based on the needs and wishes that customers may have in relation to each type of benefit.

In the next four sections, each type of pleasure is explained in a little more depth. Examples are given, demonstrating how disabilities can affect these needs and wishes. Further examples show that the requirements of the able-bodied can be equally diverse and that designing with the disabled in mind can also bring benefits for the able-bodied.

Physio-pleasure

These are pleasures that have to do with the body and senses. They include pleasures connected with touch, taste and smell as well as feelings of sensual pleasure. In the context of products physio-p would cover, for example, tactile and olfactory properties. Tactile pleasures concern holding and touching a product during interaction. This might be relevant, for example, in the context of a telephone handset or a remote control. Olfactory pleasures concern the smell of the new product. For example, the smell inside a new car may be a factor that effects how pleasurable it is for the owner.

The level of physio-p that a person gains from use of a product is likely to be affected by the fit of the product to the person's physical characteristics. For example, a product that has a number of fiddly little switches may not be pleasurable for those with limited motor control or with arthritis. It may also be displeasurable for people with long fingernails. As women tend to grow their nails longer than men, this might be something to avoid on products where women form a substantial proportion of the target group. This, then, is an example of a design decision that benefits the disabled, but has equally positive effects for another able-bodied group — those with long fingernails. Indeed, all users may prefer to avoid fiddly catches and switches.

Other examples of products that were adapted or originally designed for the physiologically disabled are the ball-point pen and the television remote control. With traditional ink pens, users without full control of their hands were often likely to break the delicate nibs. The ball-point is a far more robust design, requiring a lesser degree of motor control to operate. The remote control, meanwhile, was originally designed for those who had difficulties in moving from their chair to the television set in order to change the channel or alter the settings.

Sometimes this effect can work in reverse, with products that were designed for particular able-bodied users bringing benefits to disabled users. One example is the hands-free telephone, a product that is commonly used by the able-bodied in a number of contexts, ranging from the busy secretary in the office to the car driver who wishes to keep both hands on the wheel while driving. Again, this product can bring advantages to those who have difficulties controlling their hands. Speech technology is another area which can bring advantages for able-bodied and disabled alike. Input devices that recognise the human voice are now used in a number of computer based applications, such as word processors. These provide benefits of convenience to the able-bodied, while giving some disabled people the chance to use a type of product that would otherwise have been outside of their physical capabilities.

Speech output interfaces have also brought benefits to blind and partially sighted users. For example, Simpson (Chapter 18) describes a reading system for the blind which uses speech output as its communication medium. De Vries (Chapter 13), meanwhile, described the use of spoken feedback in a car stereo system. These two examples show how the development of a particular technological medium has created benefits for both able-bodied and disabled.

Many 'traditional' human factors issues are likely to affect the level of physio-p that a person gains from a product. For example, for furniture to be comfortable it must fit the anthropometric dimensions of those who will be using it.

Many of the traditional inclusive design issues would also fall within this category. For example, users' mobility, strength and flexibility will have design implications when considering the level of physio-p that a person will gain from a product.

Often, products can exclude people physiologically just through a basic lack of thought. For example, in a presentation to the Conference on which this book is based, Lindsey Butters of the Consumers Association highlighted a problem that many arthritic people have when using toasters. The problem arises from the resistance of the spring that forces the toast to 'pop up' when cooked. In order to lower the bread into the machine, users have to push down a lever which depresses this spring. The problem is that on many toasters this spring is too 'strong' for a substantial proportion of arthritic users. Butters pointed out, however, that the toasters would work just as well with a weaker spring.

What probably happened was that those manufacturing these toasters did not consider the needs of the weak or arthritic when specifying the design. This thoughlessness led to the needless exclusion of arthritic people from the user group and a subsequent needless reduction in the potential customer base.

Socio-pleasure

This is the enjoyment derived from relationships with others. The term 'relationships' is used in a broad way here, to mean anything connected with interaction between two or more people. Examples might be conversing with friends, enjoying time together with loved ones, or being part of a crowd at a public event. Relationships might also be seen in terms of co-operation or competition between people. For example, working with colleagues towards a common goal might be a source of pleasure. Likewise, getting the upper-hand over the 'opposition' may also be pleasurable.

Status is another social issue that can be a source of pleasure — the recognition of a person's achievements or 'standing'.

Products can facilitate social interaction in a number of ways. For example, a coffee maker provides a service which can act as a focal point for a little social gathering — a 'coffee morning'. Part of the pleasure of hosting a coffee morning may come from the efficient provision of well brewed coffee to the guests. Other products may facilitate social interaction by being talking points in themselves. A special piece of jewellery may attract comment, as may an interesting household product, such as an unusually styled TV set. Association with other types of products may indicate belonging in a social group — for example Dr. Marten's boots for skinheads. Here, the person's relationship with the product forms part of their social identity. Others may want products to reflect their social status in the way that the Porsche and the Filofax became symbols of the yuppie culture in 1980s Britain.

People's social lifestyles may also be a factor which has a major effect on the level of pleasure that they gain from a product. In particular, this is because it may affect the role that the product plays in their lives. For example, the extent to which a person entertains visitors may affect the roles that products play in their home. When others see the products that a person has in his or her home they may make assumptions about the status of the person and whether or not they have good taste. In this context, then, products, which in private might stand or fall on their functionality and quality of use, have taken on another role — that of social indicators.

Often, it seems that manufacturers expect disabled people to trade off socio-benefits for the physio-benefits that a product can bring. Whether this expectation is implicit or explicit is not a matter for discussion here. However, the results of such attitudes can be seen in some of the hideous looking products that are designed to meet disabled people's physiological needs. In the UK, two examples of products that epitomise this are the little blue three-wheeler cars specially designed for disabled drivers and the special shoes issued to people with club-feet. Both of these products draw attention to the user's disability and, because they embody aesthetics that few would accept given a choice, immediately 'label' the user as disabled and unempowered.

Often products designed for the disabled have a medical or clinical aesthetic. They are often sold primarily through catalogues and specialised stores, which can make them difficult to obtain, service and repair (Mullick and Steinfeld 1997). This may be due to the comparatively small market for these products (as compared with mass marketed products). Again, this can make the products stigmatising, as their appearance communicates the users' disabilities.

Inclusive design approaches can eliminate much of the social stigma associated with products for use by the disabled. There is no stigma attached to using a television remote control or a ball-point pen. Because these products are also suitable for the able-bodied there is a mass market for them which means that there is commercial justification for significant aesthetic design input. This gives disabled people sufficient choice to be able to express their tastes in the same way as anyone else — empowering the disabled to make choices and express themselves. Creating mass-market products that can be easily and discreetly adapted is another effective approach. For example, driving an adapted 'normal' vehicle is far less stigmatising that driving a specially designed invalid car. If someone who already owns a car becomes disabled it would be far preferable for them to be able to adapt their current vehicle, rather than have to sell it and buy an invalid car. Similarly, where there are disabled and able-bodied drivers in the same family, it may be preferable if a single vehicle could be adapted so that they could both drive it, rather than having to buy an invalid car in addition to their original vehicle.

Psycho-pleasure

Psycho-p refers to pleasures of the mind. It includes pleasures of doing and completing tasks as well as pleasures associated with particular states of mind — for example excitement or relaxation. It is the type of pleasure that traditional usability-based approaches are perhaps best suited to addressing, as these tend to address the effectiveness, efficiency and satisfaction with which tasks are completed. It might be expected, then, that a usable product would provide a greater level of psycho-p than one which was not usable. For example, a wordprocessor which facilitated quick and easy accomplishment of, say, formatting tasks would provide a higher level of psycho-p than one with which the user was likely to make many errors. Other products have been designed with the purpose of directly influencing people's moods. For example, products that emit soothing sounds or aromas.

Personality may have an effect on the design characteristics that a person finds pleasurable in a product. It might be, for example, that practically minded, 'down-to-earth' people may prefer rational design forms, while more flamboyant or imaginative people may appreciate a greater emotional component to the design. Rational approaches to design tend to encourage the creation of forms that directly and solely support the function of the product, whereas emotional approaches may encourage the inclusion of additional elements such as metaphor or humour in the design.

The psycho-p associated with products may vary according to the circumstances in which users interact with them. Sometimes little 'quirks' in a product, which may be seen as adding 'charm' when the user is relaxed and under little pressure, can be seen as major irritations when the user is stressed and hurried. For example, the handles on some of the more exotic 'designer' kettles can become very hot, requiring the user to wrap a tea-towel around the handle in order to pour. This might be seen either as adding character or as being a crass design flaw in the product, depending on the mood of the user at the time.

Inclusive design approaches recognise that different people have different levels of cognitive ability and different levels of experience with particular types of product and different attitudes towards products. For example, older people, who did not grow up surrounded by information technology (IT), may not feel as comfortable with IT products as younger people do. People with lower levels of working memory may also have difficulties interacting with products (Freudenthal 1998). Older people are likely to have lower levels of working memory than younger people, although deficits in working memory may also occur for other reasons.

Freudenthal (1998) suggests that designers should minimise the load on users' working memory by reducing the level of 'new' knowledge that is needed to operate a product. This prinicple is known as compatibility (Ravden and Johnson 1989). It means that designers should create products that draw on users' previous experiences for their operation.

The Apple Macintosh operating system interface is an example of the use of this principle. In this case the interface has been designed to be compatible with an office environment. The interface contains icons representing items such as documents, folders and trash cans. These represent, respectively, word processing files, clusters of files, and the delete function. The user can manipulate these icons using a pointer controlled by a mouse. For example, the user can move an icon from one position on the screen to another, by using the mouse to 'drag' it. If, for example, a document icon were to be moved to a trash can icon, then a word processing file would be deleted. This might be what the user would expect based on their experience of office environments — if a document is put in the trash can it is discarded. Here, then, experience gained in an everyday life domain, in this case the office, has been made useful in the product domain — in this case a personal computer.

Another useful principle, for those with either short- or long-term memory deficits, is that of 'storing knowledge in the product', rather than expecting users to store knowledge in their heads (Norman 1988). Again, this principle is utilised in the Apple Macintosh interface, which relies on users selecting commands from menus or clicking on icons in order to execute commands. This sort of interface is known as a 'direct manipulation' interface. In this case the product presents the user with a number of choices from which he or she has to make the appropriate selection. This is known as 'knowledge in the product', as the product is 'telling' the user what the options are and simply 'asking' the user to make a choice from these options. This is in contrast to the command-line interfaces that dominated the market before the Apple Macintosh was manufactured. With these interfaces, the user had to type a string of characters in order to get the computer to execute a command. For example, to duplicate a file the user might have to type "COPY" followed by the filename of the document to be copied.

Command-line interfaces rely on users storing knowledge in their heads. For example, users would have to remember that the command for duplicating documents was "COPY". They would also have to remember the filename of the document that they wanted to copy. With direct manipulation interfaces, by contrast, the user only needs to recognise the command that they need from a set of options that is presented to them.

Clearly, it might be expected that direct manipulation interfaces will bring special benefits to those with memory deficits. However, this was not the express intention behind the Apple Macintosh. The direct manipulation interface was intended to make computers easier for all people to use. Indeed, it was this ease of use that led to the commercial sucess of the Macintosh as a mass-market product. This is an example of a well designed product providing spin-off benefits for 'special' groups.

Ideo-pleasure

Ideo-p refers to the pleasures derived from 'theoretical' entities such as books, music and art. In the context of products it would relate to the match of a person's values with the values embodied in a product. These include aesthetic values — such as judgements about what is attractive — and moral values. For example, a product made from bio-degradable materials might be seen as embodying the value of environmental responsibility. This would be a potential source of ideo-p to those who are particularly concerned about environmental issues. Ideo-p would also cover the idea of products as art forms. The video cassette player that someone has in the home is not only a functional item, but something that the owner and others will see every time that they enter the room. The level of pleasure given by the VCR may be highly dependent on how it affects its environment aesthetically.

Of course people differ widely in both their aesthetic and moral values. Nevertheless, there may be a degree of systematic variation across groups. Whilst aesthetic tastes in Europe appear to be becoming more understated, Far Eastern markets are starting to become influenced by the desire for conspicuous consumption. Designs that emphasise a product's financial value may therefore be appreciated in these cultures. Similarly, Europeans' desire for the removal of what many see as spurious functionality may not be reflected in the Far East, where extra functions may still be seen as giving better value for money.

It is also possible that age may be associated with some systematic variations in values. An analysis of Western culture, originally conducted in the context of fashion design, showed that different decades have tended to be associated with different consumer values (Steele 1997). In the 1950s for example, fashion tended to emphasise conformity and status, as people clamoured to be seen as respectable members of the middle classes — the people taking the lead in returning the West to prosperity after the ravages of World War 2. The grey flannel suit was, perhaps, the archetypal male attire of the decade. By the 1960s, youth had become the most prized value. This was reflected in clothing that emphasised youthful sexiness, with the mini-skirt becoming the female fashion item of the decade, while males turned to elaborate and colourful 'peacock' attire.

The values that were important in past decades may influence the values that particular age groups hold today. People who were in their twenties or thirties in in the 1960s — people who are now in their late fifties or early sixties — may still value products that have a youthful image. In contrast, those who were in their twenties or thirties in the 1950s — people who are now in their late sixties or early seventies — may put more value on products that emphasise status. When targeting products at these groups it is important to have an understanding of the values that may be associated with each generation. Simply designing products for 'old' people is likely to be seen as patronising and betrays a lack of understanding of the user.

Once again, taking an inclusive attitude towards these age groups can have wider commercial benefits. In the context of the automobile industry, for example, sporty, youthful looking cars may appeal to people now in their late fifties and early sixties, a group that is likely to have a higher than average disposable income. Of course, this type of vehicle may also appeal to many younger groups in the population, such as 'yuppies' or others who wish to express their youth and status. The Kansei Engineering technique (Jordan, Chapter 23) was used in the development of the Mazda MX5 Miata to ensure that the car's design carried associations with youth and status (Jordan and Macdonald 1998). This should, then, make the car appealing to those in their fifties and sixties as well as those in younger age groups.

Much design for the disabled appears to ignore the values that the disabled may have. It seems to do this in two ways. Firstly, the inelegant 'medical' designs that are so often associated with this area tend to reflect a view in which the users are defined by their disability rather than their values. Just as different able-bodied people have different values and aspirations, so may disabled people exhibit a similar diversity. In particular, the association of disability and old age should not be taken for granted. For example, an elderly person who has lost the use of his or her legs through disease may have very different values from a young person who has lost the use of his or her legs in a motorcycle accident. This person has already suffered terribly as a result of their accident, why should they have to suffer further though marginalisation — the loss of choices that others in society take for granted.

It is inclusive design approaches that make it commercially viable to give choices to disabled people. Consider the case of a domestic product, such a vacuum cleaner. If manufacturers were to design vacuum cleaners that could be operated and towed from a wheelchair as well as from a standing position, then wheelchair users and the able-bodied would have an equal choice in the marketplace. This would also bring manufacturers a commercial advantage as each model of cleaner would have a bigger potential market than if it could only be used by the able-bodied.

A development in vacuum cleaner design that should prove beneficial to the disabled is the provision of controls at the top of the dust-tube, at the point where the user holds the cleaner. This saves the user from having to reach down to the main body of the vacuum cleaner in order to operate the controls. Of course, this is beneficial to all users, as it is far more convenient for users than having to bend down every time that they want to operate the controls. This is a further example of how good, usable design can benefit all and provide the economy of scale necessary to give all a choice.

The second way in which design for the disabled may ignore the values of disabled people is through a lack of understanding of some of the particular views and values that may result from being disabled. These values may result from having to confront issues and situations which are less likely to form part of the experience of an able-bodied person. Two examples of such issues are exclusion and prejudice.

On a simple practical level, issues such as lack of access and facilities can lead to disabled people being excluded from many of the everyday activities that able-bodied people take for granted, such as going to pubs and clubs and other public places. Even educational establishments, such as colleges and universities often offer poor access for the disabled. On a more 'symbolic' level, disabled people are often under-represented in media representations of society. For example, the two most watched soap operas broadcast by the BBC — *Eastenders* and *Neighbours* — currently (i.e. in May 1998) have not one single disabled character between them. It seems almost as if some sections of society would rather not even acknowledge that disabled people exist. Unfortunately, this attitude may extend to the workplace, with disabled people being under-represented in most of the professions.

Because of this exclusion and prejudice, disabled people may be particularly displeased with products that appear to label or marginalise them. Compromises in design quality and choice may particularly annoy disabled people because of this. These then, are ideological issues that arise from the social context in which disability is viewed by others. They provide further reasons why inclusive design — design that doesn't treat disabled people as a 'special' group — is particularly welcome.

17.3 CRITERIA FOR INCLUSIVE DESIGN

So far in this chapter, some of the benefits of inclusive design have been outlined and examples have been given of how inclusive design can benefit disabled and able-bodied alike. This section of the chapter looks at the criteria by which a design can be judged as being inclusive.

Seven criteria for inclusive design are outlined, along with an example to illustrate each. They are based on the definition of 'universal' design offered by Molly Story of the Center for Universal Design at the North Carolina State University (Story 1997). They are as follows:

1. *Equitable use.* This means that the design should be usable by, and marketable to, any group of users regardless of whether or not they have a disability and that a design should avoid marginalising or stigmatising any group of users. For example, buildings should offer both stair and ramp entry options, so that all can enter them.
2. *Flexibility in use.* This principle embodies the idea that a design should support a choice of different methods of usage. Story gives the example of automatic bank teller machines. These, she asserts, should offer information in visual (visual display unit), tactile (raised letters or Braille) and audible (speech) formats. They should also be reachable by tall, short and seated users. She also points out that buttons are easier to push if they are large and well spaced.
3. *Simple and intuative use.* Designs should be simple and intuitive to use, regardless of, for example, the users experience, knowledge level, or language skills. This priniciple should be applied as far as is practical. Whilst it would not be expected to apply to products such as aircraft — where it would be accepted that a pilot would need some training — it should apply in the context of 'everyday' products. These would include domestic appliances, such as stereos, TVs or kitchen appliances and systems for public usage, such as ticket machines, automatic teller machines and public information systems.
4. *Perceptible information.* As a simple example, consider the sound that a microwave oven makes when it has finished cooking. Because the cooking process is relatively quiet, the addition of an artificial sound such as a 'beep' or a 'ping' is often included to let the user know when the cooking is done. This is particularly useful for a person with a visual deficit, as it may be the only indication to him or her that the food is ready. However, this also brings benefits to other users. It means that users do not have to keep watching the microwave's display in order to be aware of when the food is ready, but can get on and do something else (watch TV, prepare something else in the kitchen) and that they only have to return to the microwave when they hear the audiable signal.
5. *Tolerance for error.* Products that are designed to avoid errors in use are beneficial to all and especially the disabled. For example, big button telephones that include 'wells' for targetting fingers, big brightly coloured buttons and large, clear labelling on the buttons are helpful in preventing the user from making dialling errors. This sort of design is particularly beneficial to people with visual deficits, or people who have motor difficulties that may make dialling difficult. However, the design brings benefits for all as it enables users to dial more quickly and with less chance of a misdial than would be the case with a conventionally designed telephone.

6. *Low physical effort.* The television remote control is an example of an adaptation to a product that has benefited everyone, but has benefited the physically disabled in particular. This product allows those with mobility deficits to interact with their TVs without having to struggle to move to the TV set itself. This product is also beneficial to the able-bodied as it allows them to control the TV from the comfort of their armchair — a more relaxing experience.

The TV remote control provides benefits for all users, irrespective of their physical abilities. Now nearly all TVs are sold with a remote. Indeed, remote controls are increasingly common in other product areas, such as audio systems and lighting.

7. *Size and space for approach and use.* Booths for public access information technology or telecommunications systems should be big enough to accomodate mobility aids such as wheelchairs. This additional room can also be pleasant for the able-bodied.

Consider, for example, a public telephone booth in a city centre. Making the booth spacious enough to provide for wheelchair access would also allow the able-bodied person to enter the booth with, for example, shopping bags. This example could be applied to other public booths, such as tourist information systems, or financial services systems.

17.4 IMPLEMENTING INCLUSIVE DESIGN

So far, it has been argued that there are many advantages to taking inclusive approaches to design — inclusive design can bring benefits both to consumers and to manufacturers. However, for manufacturing industry, inclusive design is still a comparatively new concept. While it is certainly to be hoped that, in the long run, design approaches that are inclusive will become pervasive, it should also be recognised that the likelihood of this happening will be dependent on some early successes.

In this section, the conditions under which inclusive design can most effectively be implemented are discussed, and some examples of successful inclusively designed products are given.

The success of any new approach will be dependent on six factors, see below. Steinfeld (1997) has shown how these principles apply to the implementation of inclusive design principles. They are discussed briefly below.

1. *Relative advantage.* The advantage of a new approach must be clear to those who may invest in the idea. The case for inclusive design has already been discussed at length in this chapter. In the end, the case boils down to moral and financial imperatives.
2. *Communicability.* In the first instance it is up to those who believe in an idea to communicate it clearly to those who may invest in it. There is growing evidence that this has happened for arguments for inclusive design, at least in principle. Inclusive design is being taught in design courses and is the subject of research, both in industry and academia. There are also a number of examples of inclusively designed products already on the market (see IDSA (1997) for examples of all of these).
3. *Compatability with norms.* An idea will have a greater chance of succeeding in practice if it fits in with the Zeitgeist — the spirit of the age. In this sense, inclusive design seems to be an idea whose time has come. Trend analysts are predicting that the new century will see an increase in 'compassionate capitalism'. They predict that the firms that will be successful will be those that are seen to provide products and services that satisfy real needs and which reflect responsible, caring values. The

practice of inclusive design can, as has been argued in this chapter, be seen to embody such values.

4.   *Nonpervasiveness.* The greater the number of aspects of an organisation that will be affected by a new approach, the smaller the chance that the new approach will be implemented. This may be a particularly important point to be aware of for those trying to win support for implementation of inclusive design within their organisation. In particular, any approaches that are likely to lead to re-tooling requirements are likely to meet with resistance. For example, Steinfeld (1997) reports a case where a cabinet manufacturer was unwilling to produce an adjustable cabinet — which would have brought clear benefits for wheelchair users and able-bodied people alike. The company fully recognised the benefits that the adjustable design would bring and supported inclusive design approaches in principle (indeed they had worked with a centre specialising in inclusive design in order to develop the concept in the first place). In the end, however, it became clear that it would be necessary to re-tool in order to manufacture this design. This became too time consuming and expensive to make it practical.

At the other extreme, consider the example of the toaster spring given in the talk by Butters (recounted earlier in this chapter). In this case the design could be made more inclusive simply by changing the strength of the spring. This could almost certainly be done without any knock-on effects. Indeed, the chances are that the spring would be supplied by a sub-contractor, so all the manufacturer would have to do would be to order a different specification of spring.

5.   *Reversibility.* Manufacturers are more likely to accept a new approach if they feel that any decision to implement it is reversible. It is easier and cheaper to reverse the decision to alter a single product component (for example a toaster spring) than it is to undo the re-tooling of an entire production line.

This principle also applies to the customer. Earlier, Coleman has reported on the Good Grips range of utensils — a range that has proved to be a commercial success. Steinfeld (1997) suggests that one of the reasons for the success of these products may be that they are inexpensive to purchase. The customer can afford to take a 'risk' with them, knowing that they can afford to replace them if they do not like using them.

A product which failed, at least partly on the grounds of reversibility, was a bath-tub made from soft, flexible materials. This bath-tub gave advantages over a contentional tub, because it greatly reduced the risk of injury should someone slip. Steinfeld (1997) suggests that one of main reasons that it didn't come on to the market was because, compared with products such as kitchen utensils, it would have been an expensive product to buy. It would cost further time and money to install, all of which would make it comparatively costly and time consuming to replace.

6.   *Number of gatekeepers.* The fewer the number of people between the advocate of a new approach and the customer, the more likely the approach is to be implemented. This applies both within manufacturing organisations and at the point of sale of products. Within manufacturing organisations, this issue is likely to be linked with the pervasiveness of the idea and the level in the organisation at which the inclusive design is explicitly supported.

There may be some inclusive design decisions which can be implemented by a designer acting more or less alone. This might include, for example, font sizes of labelling, the texturing of buttons and the use of colour. However, once technological or manufacturing changes to a product are required, others such as production engineers and development engineers will have to consent to such changes in order for them to be implemented successfully.

Similarly, the higher in an organisation's hierarchy the advocate of an approach is, the less permission he or she will need in order to have the approach implemented. It is, therefore, important for advocates of inclusive design to convince senior management of its merits as early as possible.

At point of sale, products that are sold in high-street retail outlets may be more effective carriers of new approaches then those that are sold through distributors, dealers and contractors. Steinfeld (1997) uses the examples of the soft bath-tub and the Good Grip kitchen utensils to illustrate this point. A problem in marketing the bath-tub was that the manufacturers would have had to convince building suppliers and plumbers of the merits of the product, before potential customers even got the opportunity to judge for themselves. The kitchen utensils, on the other hand, were simply sold through high street retail outlets, where the consumer could look at and handle the products and make their own judgements about their potential advantages.

17.5 CONCLUSIONS

In the first part of this chapter, a number of advantages for inclusive design were discussed and moral and commercial arguments were given in favour of inclusive design. Criteria for inclusive design were outlined and advice was given as to the circumstances in which inclusive design approaches had the highest chance of initial success.

In the long term, of course, it is to be hoped that inclusive approaches to design will pervade the whole range of products and services. Of course, it will not be possible to create all products so that they fit all people. Nevertheless, an approach which starts with this aim, rather than instantly dismissing the disabled out of hand or marginalising them by creating hideous looking 'medical' products, is one that should be encouraged. It is the morally right thing to do and is likely to bring financial reward to manufacturers who embrace inclusive design.

17.6 REFERENCES

Freudenthal, A., 1998. Transgenerational design of "smart products". A checklist of guidelines. In *Gerontechnology: A Sustainable Investment in the Future*, J. Graafmans, V. Taipale and N. Charness (Eds), IOS Press, Amsterdam, pp. 406-410.

IDSA, 1997. *Innovation.* Spring 1997.

Jordan, P.W., 1997. Putting the pleasure into products. *IEE Review*, November, pp. 249-252.

Jordan, P.W. and Macdonald, A.S., 1998. Pleasure and product semantics. In *Contemporary Ergonomics*, M. Hanson (Ed.), Taylor & Francis, London, pp. 264-268.

Mullick, A. and Steinfeld, E., 1997. Universal design: what it is and isn't. *Innovation.* Spring 1997, pp. 14-18.

Norman, D.A., 1988. *The Psychology of Everyday Things*. Basic Books, N.Y.

Ravden, S.J. and Johnson, G.I., 1989. *Evaluating Usability of Human Computer Interfaces: A Practical Method*. Ellis Horwood, Chichester.

Steele, V., 1997. *Fifty Years of Fashion: New Look to Now*. Yale University Press, New Haven.

Steinfeld, E., 1997. Innovation theory: how does an idea spread and grow? *Innovation*, Spring 1997, pp. 19-24.

Story, M.F., 1997. Is it universal? Seven defining criteria. *Innovation*, Spring 1997, pp. 29-32.

Tiger, L., 1992. *The Pursuit of Pleasure*. Little, Brown and Company, Boston, pp. 52-60.

CHAPTER EIGHTEEN
Personal Reader Systems for the Blind

SOME DESIGN ISSUES

P. J. SIMPSON

Department of Psychology, University of Surrey,

Guildford, Surrey, GU2 5XH, UK

18.1 INTRODUCTION

The development of inexpensive computing resources has made it possible to design and produce a range of advanced cognitive artefacts. These artefacts are devices designed to maintain, display or operate upon information in order to serve a representational function. When used, they enhance human cognitive performance (Norman 1991). Performance is enhanced by changing the nature of the task performed rather than by changing the capability of the individual user. When the informational and processing structure of the artefact and the human user are combined in the execution of a task, there results an increase in capacity reflecting the total system of the human user, the task and the artefact.

When designing such artefacts it is considered crucial to match the design, in particular the interface through which the user communicates, with the user's conceptual model relevant to the task in hand. The user model reflects prior task experience as well as that obtained while using the artefact. The user model provides the basis for the user's understanding of the behaviour of the artefact and guides its effective use for the task. This chapter considers the design of one such type of artifact — the generation of synthesised speech information for use by the visually impaired. The problems of perceiving this information are considered in relation to the design of tools which enable the user to explore the linguistic and conceptual content of the speech. It is argued that tools which are designed to explore the source representation are not matched to the task of speech perception and comprehension which use of the artefact involves. Thus there is a mismatch between the information required by the user and the informational structure within the artefact which is accessible to the user. Suggestions are made for overcoming this problem by enhancing the facilities provided by the artefact and the user's task knowledge.

18.2 DEFINING AND MATCHING THE USER MODEL, A LITTLE HISTORY

One classic example of how a user model was used to shape the design of an interface is provided by the case of the Star Interface design. Smith *et al.* (1982) have described the evolution of the Star Interface design which has provided a paradigm for graphical user interfaces in personal computers. The Star designers chose to base their design on the desktop metaphor since they were designing an office information system. Their interface gave access to objects and functions which paralleled those found in a physical office. The designers set out to exploit an equivalence between the desktop interface objects and functions, and a user's existing model which was the basis for their actions in a physical office. In practice the representation of a 'desktop' within an interface is somewhat schematic. This type of interface can support direct expression of concrete actions and commands, but not abstract generic commands (Myers 1991). The interface assumes a model of the office environment and way of working with information which may not be representative of the user's actual practice. Broadbent (1966) noted that the well ordered desktop and the well ordered mind are ideals which we aspire to but may not achieve! Thus the task of evolving a design to match the user model presents problems of definition, representativeness, and implementation. Fortunately human users learn. Therefore they can adapt to aspects of a designed artefact provided that the properties salient for its operation are accessible by being appropriately represented in the user interface.

18.3 OVERCOMING SENSORY DEFICITS

The design of office information processing systems is only one very evident example of the use of the microcomputer to create a cognitive artefact. It is also possible to use such artifacts to help the user overcome sensory deficits. In this section, the point will be illustrated in the context of aids for the hard of hearing. The rest of the chapter will then focus on how this issue applies to the design of reading aids for the visually impaired.

18.4 HEARING AIDS

Recent developments in the design of hearing aids have exploited microprocessor and digital techniques to process sound. In the strict sense described by Norman (1991) they are serving a representational function — converting 'real' sounds to electronic representations of those sounds. Sound can be processed to compensate for changes in an individual's threshold sensitivity (McAllister, Black and Waterman 1995). In this case the hearing aid enhances performance by selectively amplifying parts of the sound spectrum. In practice however, hearing loss can involve deterioration in a number of different aspects. By controlling loudness, introducing cue enhancement (see Hazan and Simpson 1998), frequency shifting, etc., it may be possible to compensate for problems of speech perception and selective attention by modifying the representation of the sound stimulus. In this way a hearing aid can be adapted to compensate for the particular hearing problems of the individual user.

When designing hearing aids, the information processing and representational operations have impact at the sensory processing level rather than at the cognitive level. It might seem that it would be easier to define an individual's sensory deficits (the 'user model') in terms of psychoacoustic measures than it is to define a 'user model' at the cognitive level. Indeed, understanding perceptual problems arising from a sensory loss, and designing effective information processing operations to overcome them, is in practice very difficult (Moore 1995). Nevertheless, understanding how a sensory deficit affects the user at a cognitive level may be central to dealing with the problem effectively. In the case of speech, the listener can use higher level linguistic and semantic knowledge to compensate for a degraded or transformed stimulus representation and thus to facilitate recognition. So the effectiveness of a hearing aid will depend on how the sound is processed and represented, whether the task involves listening in quiet or noisy conditions, and on the linguistic and conceptual resources available to the user.

18.5 READING AIDS FOR THE VISUALLY IMPAIRED

Access to the printed word is essential if an individual is to participate in a modern society. Computer based reader systems can give the blind access to text based information by transforming the text encoded in the form of ASCII characters into synthesised speech. When combined with an optical character reader, the resulting system offers visually impaired and blind individuals access to printed material which can be flexible and responsive to individual needs.

Nevertheless, these systems do not always prove popular with the intended users. A small informal observation of three users, all blind or severely visually impaired students, gave the impression that the reaction to the system was ambiguous.

The students were using the system to access books, articles and research papers as part of their degree course. The three users showed contrasting responses to the system. One valued the flexibility and independence the system offered. Another became irritated with the limitations of the system and preferred to work with audio tapes prepared by human readers. The third could work with audio tapes and Braille representations. Thus there was little incentive for this user to spend time becoming familiar with the operation of the system and the additional task that this entailed. Even on the basis of this small study it appears that the intended target group are ambivalent about the benefits that the reader system delivered. In practice even the enthusiastic user showed difficulties in comprehending concepts and arguments presented through a synthesised speech version of text.

These observations prompted an analysis of which features of synthesised speech based personal reader systems could present problems for a visually impaired user.

18.6 PROPERTIES OF SPEECH DERIVED FROM TEXT

The pathway from printed text to synthesised speech includes a number of steps. At each step there is the possibility of a problem occurring. The optical character reader system can run into difficulties with certain font designs, degraded text and multi-column layouts. Adaptive recognition procedures and verification routines working at the word level can overcome some of these problems (Evans 1995). Assuming however that there are no visual recognition and layout problems, the user of a personal reader system is presented with a synthesised speech representation of lines of printed text. The translation from text to a 'spoken' version is a complex process. This is especially true when the language concerned contains many exceptions with respect to spelling and pronunciation (Edwards 1991). In the best systems a large pronunciation dictionary and good software design can minimise this problem.

The result of these recognition and encoding operations is a synthetic speech representation of the text. The speech derived from the text will differ with respect to a number of features found in normal speech. Since it will be derived from written text, it is likely to be more grammatically complete but will also involve more complex syntactic forms than are found in normal speech.

Crystal (1987) has argued that written language and speech are the products of radically different kinds of communicative situations. One graphic illustration of this is provided by Raban's (1992) account of the change in speaking style which occurred during President Clinton's 1992 campaign. At the outset of the campaign Clinton used a complex language form involving branching sentences and qualifying clauses. He spoke about economics in a style reminiscent of a formal debate or an academic paper. He presented an account of a complex world in which there were no easy answers. It was clear to the listener that Clinton was reading pre-prepared text. Later in the campaign Clinton gave up the use of complex sentence construction and presented messages involving simplicity and certainty. By modifying the content of his message, Clinton changed its rhetorical impact. It now sounded 'natural' — the way a person would really speak. By simplifying the grammar of his sentences, Clinton made it easier for the listener to comprehend the message.

18.7 THE ROLE OF PROSODY IN THE COMPREHENSION OF SPEECH

A major feature of normal speech is the use of prosody. Prosody entails the use of stress, rhythm, timing and intonation to mark structural (syntactic) features in the speech stream and to convey meaning and intention. Synthetic speech lacks many of the prosodic features used in normal speech. In written text some prosody encoded features are expressed by punctuation while other features are expressed through sentence content. When expressed as words in synthesised speech, lacking intonation and stress, punctuation lacks the impact of the prosodic based features found in normal speech (Crystal 1987).

The lack of normal prosodic cues makes detection of the underlying linguistic structure more difficult. Wingfield and Butterworth (1984) have proposed that when speech lacks normal timing, pauses and amplitude variation or stress, the speech is perceived as if it were a list of unrelated words rather than a structured speech stream. Wingfield and Butterworth found that the lack of prosodic features markedly changed subjects' strategy with respect to the size of perceptual unit subjects attended to when asked to remember verbal material.

18.8 COMPLEXITY AND WRITING STYLE

It was noted earlier that, relative to speech, printed text exhibits a greater grammatical complexity. This claim must be qualified in the light of the fact that the writer of text can vary the grammatical complexity of textual material to fit the intended readers' reading skills. In the case of the small observation study, the personal reader system was used to encode degree level reading material. The sentence structures used in this material reflected the complexity of the descriptions and arguments included in it. But the use of relative clauses, material in parenthesis, nested sentences, and so on will pose problems of comprehension when presented as synthesised speech which lacks prosodic features to indicate punctuation and structure. Moreover, problems of lexical (word reference), structural (syntactic) and referential (pronoun) ambiguity can present further difficulties in achieving the understanding of synthesised speech encoded text.

18.9 STRATEGIES IN READING AND LISTENING

Studies of eye movements during reading suggest that when faced with a complex sentence or ambiguity, the reader will stop the normal left to right saccadic eye movements and engage in a regression to fixate earlier material (Rayner, Carlson and Frazier 1983). During these operations it is claimed that the pattern of fixation is governed by linguistic factors reflecting a need for further processing of the sentence content. Similarly the amount of time spent fixating an individual content word (nouns and verbs) is dependent on their familiarity.

In the case of synthetic speech encoded text, the listener can adopt a passive listening strategy equivalent to a succession of left to right eye movements along a line. With this strategy the listener could adjust the speaking rate of the speech synthesizer to allow sufficient time for understanding and comprehension.

However text to synthesised speech systems also allow the use of an active strategy to access speech encoded text. The user can choose to listen to the current, next or prior character, word, sentence, line, or paragraph. The cursor controlling the speech output can also be moved to the beginning and end of the line and top or bottom of the screen. Options exist for choosing between saying or spelling a word, or phonetically spelling a character to aid understanding.

18.10 CONTRASTING STRUCTURES FOR TEXT AND SPEECH

Designers of personal reader systems have foreseen problems leading to failure of recognition and understanding at the word level, and at the level of the sentence and paragraph. Thus they have provided a series of 'tools' to help the user review words, sentences and paragraphs in order to achieve comprehension.

The facilities for moving around the screen at the level of character, sentence, line and paragraph are based on the recognition of features of the text which are encoded 'naturally' in the ASCII representation — i.e. spaces, full stops, commas, colon characters, etc, as well as soft returns and hard returns.

The features available to the user to 'signpost' a directed analysis of the synthesised speech relate to the graphic structure of the original text encoded in ASCII form. But the 'features' of character, word, sentence and paragraph are indirectly related to the synthesised speech version of the original text which the user is actually trying to comprehend.

This point is clearly illustrated by the feature 'line'. It is possible to ask the personal reader system to 'say' the next, current or prior line, or 'move cursor' to the end or the beginning of the line. But this feature has doubtful significance for the user who wishes to understand a speech representation of the text. Sentences do not obligingly live on single lines!

18.11 MATCHING THE PERSPECTIVE OF THE BLIND USER

It seems that the personal reader 'tools' provided for reviewing selected parts of text expressed as synthesised speech is shaped by a graphics 'text on a VDU screen' perspective. But the user of a personal reader system could be working with a different model and from a different perspective. This model will reflect the linguistic and conceptual structure represented in the speech stream rather than graphic features and layout of the original text.

The ability to request that the system 'says' individual sentences, or to move through a synthesised speech representation to link individual sentences, will serve the 'linguistic' needs of the user concerned to comprehend the synthesised speech to some degree. But could this facility for review and analysis be improved? Would it be possible to provide the facility to request that the personal reader identify and 'say' individual clauses or phrases within a sentence? The problem would involve finding an ASCII character based feature which would define the boundaries of the clause within a sentence. Commas are a possible feature but in many instances there are no reliable character based feature. However, before worrying more about which, if any, ASCII character string could be used, it may be beneficial to review the principle involved here.

It has been suggested that when text is encoded into synthesised speech, the resulting representation will pose problems of comprehension for the user. These problems will arise from the restricted nature of the speech synthesis and therefore representation, the complexity of sentence structures, and ambiguity. Facilities should be made available to the user to allow review and analysis of the content of the synthetic speech which are analogous to regression and review in reading. It is noted that facilities to support this operation are provided in existing systems. They rely on features encoded in the ASCII text representation which identify a character, word, sentence or paragraph. However no provision is made to access individual phrases or clauses although these are distinct 'units' which have to be related to achieve a coherent interpretation.

There are systems currently available which allow review. However, with these, the user is expected to relate to the synthesised speech in terms of its original textual representation and layout rather than its synthesised speech form. Thus the 'tools' provided for analysis and review are based on an inappropriate model of how the LISTENING user relates to information presented as a synthesised speech representation.

In practice, the structure of any text will vary with the individual author's writing style and choice of sentence construction, as well as conceptual topic and content. Thus the listener's choice between selecting a previous word, sentence or sentences, or previous paragraph when facing difficulties of comprehension may reflect an exploratory heuristic decision rather than a directed search based on a clear grasp of how previous and present sentences relate together.

18.12 ENHANCING PERSONAL READER SYSTEMS

Research on the problems of generating intelligible synthetic speech representation from a textual representation which includes prosody is a very active area — see, for example, House (1997). Advanced systems which provide text to speech synthesis have to include several levels in the process (Benoit 1995). Text analysis aimed at determining the grammatical characteristics of the words and the syntactic structure of sentences are necessary for phonemic conversion and prosodic modelling. Grapheme to phoneme conversion can be based on the use of very large word dictionaries which simultaneously give the phonemic conversion, grammatical category, stress position, etc. The generation of prosodic modelling which assigns a duration and melodic contours to a sentence and its constituent syllables is governed by the syntactic structure and the prosodic conventions of a particular language/speech group. The generation process can be based on rules and/or reference profiles stored in a prosodic lexicon. The prosodic lexicons of the future may be differentiated to reflect sentence complexity and rhetorical style. All these developments will lead to marked improvements in the quality of the speech representation available to the user of text to speech systems. However problems with language comprehension are likely to remain.

It is often argued that problems of sentence complexity and ambiguity, which emerge from linguistic analysis, assume less importance when language understanding is guided by prior knowledge and expectation in a 'top down' model of language perception (Best 1995). However a top down process is only likely to be efficient when the reader/listener can utilise appropriate topic relevant knowledge held in his/her memory. When the individual is reading (listening) to learn, the relevant knowledge to support the required interpretation operations may be lacking. In this case a clear and unambiguous representation of the material is crucial.

18.13 ENHANCING THE LISTENER'S CONCEPTUAL RESOURCES

Stanovich and Cunningham (1991) have argued that the activity of reading text should be considered as involving constrained reasoning intended to achieve a coherent and consistent interpretation. The same case can be made for the user of a personal reader working with synthesised speech derived from text. Therefore we should work to lessen the interpretive problem posed by an impoverished synthetic speech representation and the structural complexity of language.

To this end, tactics used in the design of teaching texts could be employed. Care could be taken to avoid complex explanatory material. Summaries of the text could be provided to establish the scope of the topic as is often done in well designed 'teaching' texts — see, for example, Solso (1994). Information could be given about the narrative structure of the text. For example it should be possible to insert a phrase indicating a new paragraph encoded in a voice pitch different from that used for the main text. An extension of this idea could take the form of a summary of the layout of each page — if pages are being encoded — describing the number of paragraphs included on each page and the clear identification of titles when included. This kind of system implies a limited form of linguistic processing to extend the information available to the listening 'reader'. At the level of an individual sentence, it might be possible to have the option to request a summary of its structure including its constituent clauses.

These suggestions appear to be potentially of value to the blind user. They arise from the idea that a visually impaired user requires an extended set of review 'tools' and guidance to aid comprehension when attempting to make sense of a synthetic speech version of text. The exploratory and review tools should support the strategies required to achieve comprehension on the basis of incomplete speech based, linguistic and conceptual information. These facilities might be extended to include commentary on grammatical as well as the 'narrative' structure of text.

Computer based systems able to classify the linguistic structure and content of text are available, see Winograd (1984) and Gazdar (1993). However systems able generate a commentary on linguistic and conceptual structure for the blind user will not be available in the near future. This is a task for human expertise. Deciding what information and commentary to provide as an aid to the comprehension of degraded speech brings into focus the question of how far an integrated and implicit perceptual activity can be guided and 'instructed'. Users of personal reader systems will have evolved strategies to aid the perception and comprehension of everyday 'normal' speech. An understanding of what these strategies are and when they are used should provide a more appropriate framework to guide the development of personal reader interface facilities.

18.14 CONCLUSIONS

Access to the written word may be seen as a pre-requisite of taking a full part in a modern society. Personal reader systems offer a way for people with severe visual deficits to have independent access to the printed page. However, in order for these systems to be optimally effective, there are a number of issues that the designers of such systems must take into account. Some of these have been discussed in this chapter — prosidy is very important to the comprehension of speech (whether synthesised or otherwise). Written text tends to be structured with far greater complexity than the spoken word. It may be more difficult to comprehend speech that is structured in the form of text. People tend to listen in a 'linear' manner, whereas with reading people may go back and forth in the text and fixate on certain words. This means that the user of a reader system may not have the same comprehension strategies open to them as when reading text.

A number of examples and suggestions were given, indicating how some of these difficulties may be overcome. Broadly, two angles of approach were suggested for addressing these issues — enhancing the reader systems themselves and adapting text so that it is suitable for conversion to synthetic speech.

18.15 REFERENCES

Benoit, C., 1995. Speech synthesis: present and future. In *European Studies in Phonetics and Speech Communication*, G. Bloothooft, V. Hazan, D. Huber and J. Llisterri (Eds), OTS Publications, Utrecht, pp. 119-122.

Best, J.B., 1995. *Cognitive Psychology*. 4th edn, West Publishing Co, Minneapolis/St Paul.

Broadbent, D.E., 1966. The well-ordered mind. *American Education Research Journal*, **3**, pp. 281-295.

Crystal, D., 1987. *The Cambridge Encyclopedia of Language*. Cambridge University Press, Cambridge.

Edwards, A.D.N., 1991. *Speech Synthesis Technology for Disabled People*. Paul Chapman Publishing Ltd.

Evans, J., 1995. *Personal Computer Magazine*. August.

Gazdar, G., 1993. The handling of natural language. In *The Simulation of Human Intelligence*, D. Broadbent (Ed.), Basil Blackwell Ltd, pp. 151-177.

Hazan, V. and Simpson, A.M., 1998. The effect of cue-enhancement on the intelligibility of nonsense word and sentence material presented in noise. *Speech Communication*, **24**, pp. 211-226.

House, J., 1997. An integrated prosodic approach to device-independent, natural-sounding speech synthesis. In *UCL Department of Phonetics and Linguistics*. http://www.phon.ucl.ac.uk.

McAllister, H.G., Black, N.D. and Waterman, N., 1995. Hearing aids — a development with digital signal processing devices. *Computing and Control Engineering Journal*, **6**, pp. 283-291.

Moore, B.C.J., 1995. 'The shape of sounds to come'. *MRC News*, Winter 1995, pp. 32-35.

Myers, D.A., 1991. Demonstrational Interfaces: A step beyond direct manipulation. In *People and Computers VI, Proceedings of the HCI '91 Conference*, D. Diaper and N. Hammond (Eds), Cambridge University Press, Cambridge, for The British Computer Society, pp. 11-30.

Norman, D.A, 1991. Cognitive artifacts. In *Designing Interaction: Psychology of the Human Computer Interface*, J.M. Carroll (Ed.), Cambridge University Press, Cambridge, pp. 17-38.

Raban, J., 1992. 'The making of the candidate'. *The Independent on Sunday*, 6 September, 1992, pp. 6-13.

Rayner, K., Carlson, M. and Frazier, L., 1983. The interaction of syntax and semantics during sentence processing: Eye movements in the analysis of semantically biased sentences. *Journal of Verbal Learning and Verbal Behaviour*, **22**, pp. 358-374.

Smith, D.C., Irby, C., Kimball, R., Verplank, B. and Harslem, E., 1983. Designing the Star User Interface. In *Integrated Computing Systems*, P. Degano and Sandewall (Eds), North Holland Publishing Company, Amsterdam, pp. 297-313.

Solso, R.L., 1994. *Cognitive Psychology*. Allyn and Bacon Inc.

Stanovich, K.E., and Cunningham, A.E., 1991. Reading as constrained reasoning. In *Complex Problem Solving: Principles and Mechanisms*, R.J. Sternberg and P.A. French (Eds), Lawrence Erlbaum Associates, Hove and London, pp. 3-60.

Wingfield, A. and Butterworth, B., 1984. Running memory for sentences and parts of sentences: syntactic parsing as a control function in working memory. In *Attention and Performance X*, H. Bouma and D.G. Bouwhuis (Eds), Lawrence Erlbaum Associates, London, pp. 351-363.

Winograd, T., 1984. Computer software for working with language. *Scientific American*, September, **251**, 3, pp. 90-101.

CHAPTER NINETEEN

Designing Domestic Appliances for Everyone

LINDSEY ETCHELL

Research Institute for Consumer Affairs, 2 Marylebone Road,

London NW1 4DF, UK

19.1 INTRODUCTION

Consumer organisations exist to empower people to make informed consumer choices and decisions. To this end, they research and regularly publish reports on comparative assessments of domestic products — advising consumers on how to choose and which models are best.

Consumers' Association (CA), the British consumer organisation, has been publishing comparative reports for the last 40 years in its magazine *Which?* and is a well-known and well-respected source of independent and unbiased information for consumers. *Which?* comparative assessments of products include both performance testing and usability assessments. Over 30 years ago CA founded RICA, the Research Institute for Consumer Affairs. It also carries out research and publishes unbiased information, but RICA concentrates on the needs of consumers who are elderly or disabled.

In recent years CA and RICA have collaborated in developing a method of assessing domestic appliances — from washing machines to electric kettles — for their usability for all consumers, whether able-bodied or disabled. To date the results of our assessments reveal product after product with a mixture of easy to use and difficult to use features. This suggests that currently usability by all — when it is present in a domestic appliance — is there by chance rather than by design.

19.2 POPULATION TRENDS

Product designers and manufacturers should surely by now be aware that the population is ageing, and that the incidence of disability increases with age. Therefore, the potential market for domestic appliances is changing.

There are over six million adults with some level of disability in Britain alone — 14% of British adults (Office of Population Censuses and Surveys 1988). This proportion is similar across Western Europe (Sandhu and Wood 1990). Analysis of the British figures by age shows that the majority of disabled people are elderly, and also that the severity of disability increases with age. The most prevalent disability is caused by arthritis — stiff, painful, weak joints.

Like most of Western Europe, the United Kingdom has an ageing population. In 1961 under 12% of people were aged 65 or over; in 1994 this reached 16% and by 2031 those aged over 65 are predicted to be 22% of the population (Central Statistical Office 1996). It should also be noted that 90% of elderly Britons continue living in their own homes — not in residential or sheltered housing (Age Concern England 1992).

These figures indicate that increasingly large numbers of elderly and disabled people will need domestic appliances that they can use easily, in order to continue to live at home independently. RICA and CA, as providers of information to consumers, have responded to the challenge of providing comparative information on products that is relevant to the needs of all consumers.

19.3 PROJECT DEVELOPMENT

Development of the usability assessment method took place in three stages. The product features that help or hinder a disabled person when using any domestic appliance are best identified by disabled people themselves. RICA's work in this area therefore began with the commissioning of product-specific user trials, to gather data about the experiences of representative consumers. Groups of 20 to 30 people with different disabilities were observed using each of a number of brands of each product, and they then answered questions on their ease of use. Depending on the product, the users included blind and partially-sighted people, wheelchair users, ambulant people with walking difficulties, people with dexterity, coordination, reaching and strength limitations, and generally frail, elderly people. Over a period of some years, the products covered were washing machines, cookers, vacuum cleaners, microwave ovens, electric kettles, toasters, irons, food mixers and a range of other food preparation machines.

Information from the user trials was combined with ergonomics guidelines and the results of other disability research to develop *special needs convenience checklists*. The checklists were designed to be administered by laboratory staff, trained to comparatively assess products for use by disabled people. RICA's first project on special needs assessments was specifically designed to answer the question of whether able-bodied laboratory staff could assess ease of use from the points of view of disabled people. It was concluded that this is possible provided that initial work is carried out with disabled users, that staff are properly trained on how people with different disabilities use the appliances, and that the checklists are tightly structured and contain appropriate guidance. Our current training sessions include demonstrations by disabled people of the problems they have with products, and the techniques they use to overcome them. Any major changes in product design also require the involvement of users.

In recognition of the fact that what is usable by a disabled person is often easier to use by an able-bodied person, the special needs checklists are gradually being extended, so that they can be used to evaluate appliances for their ease of use by any user, whatever their ability. Justification for this approach has been shown in recent work on microwave ovens. Changes in microwave oven design and usage required new user trials. Three user trials were set up by CA and RICA, with able-bodied people, partially-sighted people and people with poor grip. Statistical analysis of the results showed that oven features which caused difficulty for one or other of the disabled groups, were also rated as inconvenient by the able-bodied users.

This method has been shown to be appropriate for a range of domestic appliances. However it is not assumed that it can be applied to any consumer product. Much depends on the task-critical nature of the product. Guidance and training will enable a non-disabled tester to assess, say, a washing machine control for use by those with poor grip. A large dial with a knurled edge is going to be easier to turn than one which is small and slippery. However, in a project on garden spades, for example, it is likely to be much more difficult for an able-bodied tester to judge the usability of a spade for someone with a back problem. Those products for which a checklist is deemed inappropriate are likely to require a special needs user trial on each project.

During the development of the methodology, RICA has liased with other research organisations in Europe working in this field. The Swedish consumer organisation, Konsumentverket (KV), has done most. In collaboration with Nordic partners, KV has developed objective methods of evaluating the ease of use of a range of appliances (Nordic Council of Ministers 1991). Based on the results of user trials, their methods involve rating product features for compliance with established, acceptable dimensions and force measurements. Such methods have the advantage over subjective assessments in that they place less reliance on in-depth training of testing staff. Objective acceptance criteria remove the need for expert judgment, thus increasing test retest reliability. However subjective assessments are practical assessments and are more closely related to the tasks carried out by consumers. RICA and CA's work in this area has shown that, with appropriate guidance (often including acceptance criteria), a well designed subjective checklist will ensure reliable and robust results. In fact Konsumentverket's findings on particular product groups are remarkably similar to RICA's, despite the different methodologies.

19.4 ASSESSMENT RESULTS

RICA produced an introductory report on a range of domestic appliances in the August 1996 issue of *Which?* magazine, assessing their compliance with design for all principles (Consumers' Association 1996). The products included washing machines, microwave ovens, vacuum cleaners, irons, toasters and food preparation machines. The results showed that almost all the brands covered had a mixture of features which would help and hinder elderly and disabled consumers. We could not recommend a single one of the 48 products assessed for their usability without reservation or qualification.

The following are summaries of the good and bad design points found on just one model of each product type.

Washing machine

For: Good grip on detergent dispenser, opened smoothly; most labelling easy to read with good colour contrast against background.
Against: Poor grip on door handle; force needed to open and close the door; door seal restricted loading space; poor grip on program and temperature controls and stiff to turn; push button controls small and close together.

Microwave oven

For: Rotary controls with a notable stop at each power setting; well spaced markings.
Against: Difficult to see food inside oven; door control and time control needed force; oven awkward to clean; turntable awkward to handle.

Vacuum cleaner

For: Comfortable gripping area on hose; automatic adjustment for carpet height; good colour contrast on labels.

Against: Extension tubes awkward to fit together; fiddly variable suction dial; small stiff manual suction slider.

Iron

For: Lightweight for dry ironing; easy to fill with water; well-spaced temperature setting markers; large dial with a ridged edge.
Against: Short handle, uncomfortable to hold; limited space for fingers beneath the handle; difficult to see water level.

Toaster

For: Wide loading lever with good grip needing little pressure; defrost and reheat buttons had small lights; good colour contrast.
Against: Buttons were small and needed pressure; crumb tray was stiff to pull out and push in; recess around loading slots.

Food mixer

For: Required little setting up of separate parts; large and well contrasted labelling; bowl and unit easy to clean.
Against: No cord storage; two hands and force needed to attach and remove the beaters; stiff control button to raise and lower mixer head; noisy in operation.

This initial report in 1996 has been followed by a substantial and on-going body of work. Usability assessments for all consumers have been carried out by Consumers' Association staff, and brief summary results published in *Which?* magazine in their regular product reports, such as washing machines and microwave ovens. At the same time, and from the same assessments, RICA has produced detailed reports as illustrated buying guides for elderly and disabled consumers (Research Institute for Consumer Affairs 1997, 1998). We continue to find and report on domestic appliance brands that have both easy to use and difficult to use features, products whose manufacturers are largely failing to create inclusive designs.

19.5 DESIGN FOR ALL PRINCIPLES

The following principles of inclusive design are simple. If followed by appliance designers and manufacturers, their products would be easier to use for all consumers.

Controls

- Should be placed where they are easy to reach without the need to bend or stretch.
- Protruding is preferable to recessed. The open space around a protruding control allows it to be operated by swollen and inflexible fingers, or even the side of a very weak or painful hand.
- Recessed controls should be avoided since there is unlikely to be enough space for a hand and they require a specific grip with the fingers.

- Well-spaced rather than cramped — together to provide enough space for swollen fingers or a hand with tremors.
- A reasonable size with a non-slippery surface. The surface should be knurled to provide friction: the greater the friction of the control surface, the less force needed to retain the grip.
- Small fiddly controls should be avoided since they put under strain the joints used in bending fingers and hands.
- The control should operate with very light pressure or force — to decrease the pressure required from the user's fingers.
- Avoid controls requiring dual action, such as push and turn, since they require continued pressure and twisting at the wrist (painful or impossible for some).
- Large clear labelling in a strong colour that contrasts against the background colour.
- Easy to find and feel tactile markings, to guide partially-sighted and blind people.
- Lights and sounds to give the user additional information.
- Touch pads should be raised (for easy location) and light to push.
- Touch pads should be in different shapes and sizes to tell them apart easily.

Stability

- Effective anti-slip bases are particularly helpful for one-handed use.
- Products that do not easily tip over.

Portable products

- No heavier than necessary to lift.
- Good balance and shape.
- Comfortable and well placed carrying handle.

Accessories

- Easy to connect and disconnect without the need for precision or strength.
- Separate storage section from the appliance to reduce its weight.
- Avoid fiddly parts which are awkward to set up.

Cleaning

- Avoid unnecessary markings and nooks and crannies that trap dust and food debris.
- Avoid sharp edges that are painful to touch or grip.

Hot surfaces

- Unacceptable where people might touch, hold or lean for support.

Instructions

- Clear, straightforward layout.

- Large typesize (14 point where possible) and good space between the lines.
- Very strong colour printing on very light colour paper.
- Matt not shiny paper — reflective surfaces can be difficult for people with impaired vision.

19.6 EFFECTING CHANGE

Domestic appliances should make household tasks easier and less time-consuming than these tasks would be without them. Domestic appliances are a tool to independent living.

There will always be a need for special aids, adaptations and individual solutions to assist the most severely disabled people with household tasks. However domestic appliances, bought as standard in the high street, should be designed to be easily used by the majority of elderly people and younger disabled people — those with the lower levels of disability. Why should our choice of usable products be reduced as we get older and weaker, less mobile and not able to see or hear so well? Elderly people generally don't think of themselves as disabled — but they do have real problems with products.

The ability to use standard domestic appliances has tremendous advantages for the elderly or disabled consumer: more choice in the high street; more competitive prices than special equipment will ever achieve; less dependence on family and carers; the good feeling that their home and its contents looks like other people's.

Greater demands are being placed on domestic appliance manufacturers as more people are living longer at home. Surely the time has come for manufacturers to respond to the challenge, by taking on board design for all principles, in the knowledge of increased market potential. The mixture of good and bad features identified by CA and RICA's work shows that easy to use features are achievable and within current technological capabilities. Freudenthal, in the following chapter, describes research which complements the investigations described here by developing design guidelines dealing more particularly with perception and cognition.

There is a need to bring about change, and organisations like CA and RICA are in a position to influence both consumers and industry. Our reports encourage people to become more discerning and demanding about easy to use products. Domestic appliance manufacturers are aware of this influence, and the effect it can have on their sales figures. They often respond to criticism in influential magazines like *Which?* and its counterparts across the world, and they have been known over time to develop product specifications that comply with consumer organisations' criteria.

There is also a need to raise awareness among designers during their training, so that using design for all principles, to produce inclusive designs for products, becomes their standard practice, rather than an optional consideration.

19.7 REFERENCES

Age Concern England, November 1992. *Housing and Older People*, **1**, Population trends and housing, England.

Central Statistical Office, 1996. *Social Trends*, **26**, HMSO, UK.

Consumers' Association, 1996. Design of domestic appliances. *Which?* August, pp. 38-43.

Nordic Council of Ministers, 1991. *Methods of Measuring the Ergonomic Properties of Domestic Appliances to Functionally Disadvantaged Persons*, Denmark.

Office of Population Censuses and Surveys, 1988. The prevalence of disability among adults. *OPCS Surveys of Disability in Great Britain*, Report 1, HMSO, UK.

Research Institute for Consumer Affairs, 1997, 1998. *Domestic Appliance Series on Choosing Domestic Appliances that are Easy to Use*, England. Separate booklets currently available on: washing machines, vacuum cleaners, electric kettles, tumble driers and microwave ovens.

Sandhu, J.S. and Wood, T., 1990. *Demography and Market Sector Analysis of People with Special Needs in Thirteen European Countries: A Report on Telecommunication Usability Issues*. Special Needs Research Unit, UK.

CHAPTER TWENTY

Transgenerational Guidelines

ADINDA FREUDENTHAL

Faculty of Industrial Design Engineering, Delft University of Technology,

Jaffalaan 9, 2628 BX Delft, The Netherlands

20.1 INTRODUCTION

"What is good for the old is good for the young," is the well-known assumption in transgenerational design. The feeling is that better products for the old can be developed by applying specific ergonomic guidelines that take into account the diminishing human capacities of the elderly, and that this will automatically provide products that are more usable for the young. If transgenerational design were to be applied in new home appliances and consumer electronics, not only would older users finally be able to program their own video recorder and use more functions on their microwave oven; but also younger users would be provided with more ease of use.

However, it may in certain cases be true that if a product properly fits the diminished human capacities and specific interests of the old it might become undesirable for the young, because it is boring to use or it looks ugly and stigmatising. It is even possible that the strategies of use of younger consumers, or their knowledge of modern apparatus, differ so much from those of older people that some guidelines for the old might be inappropriate for the young.

New design guidelines, meant for domestic appliances and consumer electronics for the general consumer market, should take the specific needs of elderly users into account, but should also benefit the young. A scientific approach in the development of such transgenerational guidelines is required. This chapter will focus on a project at Delft University of Technology aimed at such guidelines. First the state-of-the-art in industrial design for older consumers was assessed. Then new, needed, design guidelines were developed.

20.2 STATE-OF-THE-ART

A theory on domestic product use, aiming at the regular, non-professional, user is generally lacking. Not much is known about cognitive capacities when it comes to product use. Classical data from cognitive psychology do not teach us much. Isolated capacities of memory have been researched but not in the context of apparatus use. Ergonomics does not tell us about habits of users or about expectations they have about modern devices, while habits and expectations do influence product use.

When it comes to special user groups, such as senior citizens, the situation is even worse. We do know that elderly users have less capacity of memory, but we do not know how much it is reduced and what the implications of this is for product use, and we know hardly anything about compensatory mechanisms, for instance by experiences with devices. Maybe elderly users are not even aided by their experiences, but are rather hampered by outdated experiences and have the very difficult task of unlearning certain things.

Marketing, cultural anthropology and classical ergonomics do not provide many answers when it comes to questions like "how do I present dozens of functions to an elderly lady?" or "will the users understand this particular message on the screen?".

Because industry cannot sit back and wait for science to come up with all the answers, some measures are taken to come up with design requirements in specific innovation projects. One of these measures is user trials in the design process. User trials are observations of subjects representing users. These subjects are observed testing products, or models of products, in actual use. The research aims to identify problems of use and come up with information to improve the design. The method has proven a powerful tool to increase the functional quality of products.

It is a method of trial and error and it has provided us with an impressive number of tested baths, walking frames, remote controls, electric clocks and so on. We can learn from these experiences, and gradually some understanding will grow about how to anticipate recurring human behaviour.

All of these isolated cases tend to generate some additional understanding of these issues, but they do not provide much specific information, which is transferable to other products to be designed. What can we learn from the tests with a TV remote control if we are designing a telephone? We cannot implement every recommendation applying to the remote in the design of a phone. We must, for instance, know which of the recommendations apply for general human behaviour, which apply for components also included in the phone and which apply for specific aspects only related to the TV remote and not to the telephone. These insights are generally lacking.

It would be much more practical if there were a theory that predicts domestic product use, for various user groups, such as the elderly. Only this could prevent designers from repeating the same mistakes.

20.3 RESEARCH AIM

The aim of our research is to generate a checklist of design guidelines for designers, concerning domestic apparatus. These guidelines should warn the product developer of 'flaws' in design that are now present in many products and cause problems during use for a substantial part of the user population. Especially elderly users encounter many problems and constitute a substantial portion of the target group of regular home appliances and consumer electronics; therefore the elderly should explicitly be included into the investigation.

Approach

We chose usability trials as a method for developing guidelines for designers. The aim, however, was not as usual one of defining usability problems of a specific product, but to identify problems that generally occur within the product classes of domestic appliances and consumer electronics, and for a substantial portion of the users, for instance, for older users.

First we observed senior citizens in their homes using their own apparatus in a close-to-natural way. From the problems and general behaviour observed, we generated preliminary guidelines that were subsequently tested as hypotheses in further observations with various age groups and in product development projects in industry. Then a final list of transgenerational design guidelines was generated.

User trials of domestic devices in the homes of elderly users

In our experiments we found some very clear problems recurring in several of the apparatus, for example, the visual aspects were often poorly designed. One of the hi-fi products we tested had black labels on black knobs, and several TVs had very small knobs, labels and icons, see Figure 20.1.

However, much of what happened was related to cognitive aspects. Users did inexplicable things, at least at first sight. It seemed that users did not randomly do non-logical things, but rather that they had knowledge in their minds that was not known to the researcher, i.e. the users' mental model. To be able to understand what the subjects did, these mental models needed to be assessed first. With that extra knowledge about the information used by the subjects to decide what to do next, the usability problems could be analysed. The problems that occurred frequently in several products and for several users were selected and guidelines were generated that indicated what measures could be taken to prevent these problems from happening. All measures were recommendations on changes in product design (e.g. a guideline can be "Do not provide more than one label with one knob" and not, for instance, "Provide better education to the consumer").

Many of the preliminary guidelines that were developed aimed at the design of clear, understandable and consistent information provided by the product and its manual. But also guidelines on other aspects were generated, such as on design methodology.

Figure 20.1 During one of the trials in the homes of senior citizens, the subject needed extra lighting to be able to see the icons on the small knobs needed to program his TV. Because he also had to hold his manual and operate the controls an extra hand from the experimenter was necessary to hold the lamp (Freudenthal 1999: 101).

20.4 TESTING THE PRELIMINARY GUIDELINES

Two actions were taken to test the preliminary guidelines. These were tested with observations of use with subjects of various ages. Besides this, the usefulness and usability of the guidelines were tested in innovation projects in industry. There were two parallel lines of research:

Observation research

Predictions of usability problems of a more complex apparatus, a TV/VCR, were generated, using the preliminary guidelines.

Twenty subjects of various ages were observed using the set, which was newly on the market. The video registration showed problems of use. These were compared to the problems predicted.

To understand these problems, inference of mental models was not necessary (as with the first experiment). The subjects were learning gradually at an easy pace to follow, so it was clear how internal information was building up and it was also clear what aspects in the product were related to the problems observed.

We could not only test the preliminary guidelines as hypotheses, but could also extend and specify them further and we could assess which guideline applied to which user group.

Figure 20.2 A design proposal for an audio system for the 'mature market'. Only the remote control of the designed system is shown. It was designed by Luuk Platschorre for the Human Behaviour Research Centre and Audio Systems of Philips Corporate Design. In this project the two interim reports with the literature study and with the preliminary guidelines were used (Photograph by Royal Philips Electronics).

Product innovation projects

The projects in industry had two main aims:

- to assess what information designers actually need in innovation projects aimed at user-centred design, including older users in the target market, and
- to test the usefulness and usability of the preliminary guidelines and their effectiveness on the quality of the final designs.

Design information relevant to design for the elderly, derived from literature and presented in a report, was tested in twelve projects of product development in industry. In four of these projects the preliminary guidelines were also tested.

To provide answers to our research questions interviews were held with the designers after the projects were finished.

In Figure 20.2 an example from one of the product development projects is shown.

20.5 CONCLUSIONS FROM THE OBSERVATIONS WITH THE TV/VCR

Practically all predictions, based on the preliminary guidelines, about what problems were to be expected with the older subjects were substantiated by problems observed. In general it was confirmed that younger users have the same strategies and approaches as elderly users. Our general conclusion therefore is that almost all design guidelines apply for young and old consumers. Some of the guidelines needed to be specified for age groups in the final list, because some age related differences were found.

Problems observed

To give an impression of observed usability problems with young and old subjects, some examples will be given. The chosen examples are all about one aspect of the apparatus, the menus on the TV/VCR screen and related control buttons. In Figure 20.3 an example of such a menu is shown. The items in the menu could be set by using the remote control. First some problems that occurred for all ages will be discussed and then the cohort-specific problems will be presented.

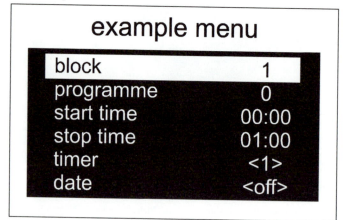

Figure 20.3 A menu comparable to some of the menus that could appear on screen is shown. A highlighted bar indicates the item that can be changed at the present time.

Problems with menu control for users of all ages

There were several inconsistencies in the test-product that caused problems for all users of the apparatus. Most item-settings in menus needed confirmation by pressing the key labelled 'OK'. This was remembered as the general rule by most subjects. They had severe problems with inconsistencies. In one case 'OK' activated a channel-search instead of confirming it to be set. All users frequently made errors activating 'search' after the search had already been completed. Another inconsistency was that two items had to be changed in a different way from all the others. They could not be changed efficiently by the cursor buttons, but had to be changed by the digit-keys. Young subjects were especially hindered by the sudden lack of 'cursor-control'.

All users were hampered by a lack of information about what was required while programming a list of settings. They did not know what had been programmed already, what was still vacant and what possibly could still be added. The menu design did not provide such information. This aspect, together with the fact that settings could not be checked during the procedure and that mistakes could only be corrected by starting over again, caused all subjects to erase (several) previously programmed stations.

When an item setting was confirmed, the cursor immediately jumped to the next item. The attention of all users was directed to following the cursor and therefore usually not to checking the last item. Mistakes (for instance forgetting 'OK') were therefore often not noticed. This problem was even more prevalent for younger persons.

Problems with menu control that depended on age

Most users can work well with menus on the screen (Figure 20.3), but about half of the elderly subjects (aged more than 59) did not know the principle; three out of ten did not even know they should look at the TV-screen to find important information. The manual did not mention this either. Some elderly subjects did not learn to use the screen as an information provider during the whole experiment. The highlighted bar over the active item did not provoke any relevant reaction.

Activating or deactivating the menu (by pushing a menu-button on the remote control) caused problems for half the elderly subjects. They did not know what to do. All subjects between 30 and 40 years of age and 50% of the elderly subjects had problems finding out how to change an item (this should be done by use of the cursor-buttons on the remote control, formed as triangles). These aspects seemed not to be communicated by the information on the product.

Younger persons (15-18) did not have these problems. Apparently, they knew the principles or they were able easily to understand them.

20.6 THE GUIDELINES

A final list of about a hundred recommendations for product design and for the design of product manuals was compiled. Backgrounds, based on empirical findings from our own research and on information from literature, for instance about changes in cognitive capacities when growing older, were related to the relevant guidelines.

The topics in the design guidelines are: necessary functionality and the ways to organise and present this to users, aesthetics (e.g. prevention of stigma), users' expectations, cognitive aspects of the presentation of information by the product, aspects of perception, and physical aspects. Besides this, recommendations are given on methodological issues (e.g. about how to conduct user trials as a design tool in an innovation project).

According to the designers who tested the preliminary guidelines they were particularly powerful when used as a set; therefore lists with results that might suggest some sort of completeness will not be presented here. Nevertheless, to provide an impression of the guidelines, some examples related to menu design will be given in Table 20.1 (on the last page of this chapter). The complete set will be available soon.

What designers need soon

We have seen that it is possible to generate design guidelines that have a predictive value (and are as such mini-theories on interaction between users and product-characteristics). Usability problems could be anticipated based on the guidelines developed. This presents some hope that the generation of a broader predictive theory on product use is not as far away as it seems. However, we will first step back to industrial reality, describe the present situation and indicate what designers really need quickly in order to improve their designs.

What we see with manufacturers of consumer durables, is the application of usability trials in different phases of product innovation as a makeshift measure to compensate for a total lack of theoretical backup. This is especially the case for design of consumer durables for the elderly. Domestic apparatus and consumer electronics that have an extreme low level of usability for (older) users can be redesigned, by the application of user trials, into products that have a higher level of usability and this is indeed happening in some instances. This should not give us the impression that these improvements are satisfactory.

Better results should be possible by founding first concepts on basic design-relevant knowledge. However, designers who try to trace relevant design information for their design projects find only a few relevant data on human capacities and design guidelines, and these are scattered across various papers in various scientific fields and presented in scientists' jargon. The designers interviewed in our project, indicated that most of the available guidelines in literature are very general, and therefore useful in an early phase of innovation, or they are very specific, and therefore more useful to detail concepts in the last phases of product development. Very few guidelines are available that can help define the basic properties of a product concept for domestic apparatus for a target market including the elderly.

In conclusion, we can say that what is really needed now are data on:

- quantitative facts about the diversity of physical, sensory and cognitive capacities, as far as relevant to product use;
- insight into cognitive processes, as far as is relevant to product use;
- more guidelines aimed at a general market, including older consumers;
- and, eventually, a broader theory about product use with, at least some, predictive value.

The expectation is that our final guidelines will meet some of these needs. The preliminary guidelines were found especially useful for the definition of the right concept and for the definition of the right properties of the interfaces of apparatus for non-professional use. The guidelines cover a broad variety of relevant aspects such as cognitive issues, aesthetics, human habits and so forth, and also provide design methodological recommendations. They are presented as a checklist for designers, who do not have much time to conduct lengthy literature searches. In the prepared volume backgrounds are provided, especially about cognitive and sensory aspects, from literature, and also from our own empirical research.

As long as sufficiently comprehensive data on human capacities and product use are still lacking, further development of usability trialling as a method will remain important. This should be done in parallel with research efforts in generating backgrounds for design, aiming finally at a predictive theory on product use.

Table 20.1 An impression of a few of the guidelines. They apply to young and old users, unless otherwise indicated.

Examples of guidelines

Some tested and adapted guidelines, relevant to menu-design are:

- For all ages feedforward as well as feedback should be complete (all users' actions should be supported, even the most obvious) and should be consistent (everywhere and every time of the same type, even in details).
- Information on static products (such as TV sets) should be substantially enlarged for elderly users, and often the contrast-rate should be higher for them.
- Be careful using icons because their meaning is, apart from a few exceptions, not understood by elderly people at all. Younger people don't always know the meaning of icons either. If provided as search cues they are usable for all ages, provided that they are clearly distinct from each other, certainly in their main graphic elements.
- Use familiar language, non-technical, no (foreign) abbreviations and avoid 'techno speak' (mainly English shortened words that stem from a technological origin). For elderly users their native language should be used.

Some new guidelines, derived from the TV/VCR experience, relevant to menu-use are:

- The chosen form of feedforward may not distract the user.
- It should always be made clear to the user in what step of a procedure he or she is.
- Information should be provided about what settings are needed, what the possibilities are and what has been done already.
- It should always be possible to check settings and to correct mistakes, even during procedures, and the user should be guided on how this should be done.

20.7 REFERENCES

Freudenthal, A., 1999. *The Design of Home Appliances for Young and Old Consumers.* PhD thesis, Delft University Press, p. 101.

CHAPTER TWENTY-ONE

Pleasure with Products: Human Factors for Body, Mind and Soul

PATRICK W. JORDAN

Philips Corporate Design, Building W, Damsterdiep 267,

P.O. Box 225, 9700 AE Groningen, The Netherlands

21.1 INTRODUCTION

Humans always have and always will seek pleasure. The artefacts and products with which we surround ourselves are potential sources of pleasure. The role and methodologies of the human factors profession make the discipline the natural vehicle for assuring that products are designed in such a way that they are pleasurable for those who own, use and experience them.

This chapter describes and defines the concept of pleasure with products. This is explained in the context of a hierarchy of user needs. A framework in which to consider pleasure issues is given and the challenges that human factors faces in order to assure product pleasurability are discussed.

21.2 IN SEARCH OF PLEASURE

Since the beginning of time humans have sought pleasure. We have gained pleasure from the natural environment. From the beauty of flowers or the feeling of the sun on our skin. From bathing in soothing waters or the refreshment of a cool breeze. We have actively sought pleasure, creating activities and pastimes to stretch our mental and physical capabilities or to express our creative capabilities. Cave dwellers wrestled to test their strength and expressed themselves by painting on the walls of their dwellings. Today we 'pump iron' in the gymnasium and decorate our homes with selections of paintings and posters.

Another source of pleasure has been the artefacts with which we have surrounded ourselves. For centuries humans have sought to create functional and decorative artefacts; artefacts that have increased the quality of life and brought pleasure to the owners and users. Originally, these objects would have been clumsily bashed out from stone, bronze or iron by unskilled people who simply wanted to make something for their own use. As systems of trade and barter were developed specialist craftspeople emerged, creating artefacts for use by others in the community. Today, most of the artefacts that we surround ourselves with were created by industry.

21.3 DESIGN AND HUMAN FACTORS IN PRODUCT CREATION

The product creation process will differ from product to product. However, it is probably a truism that design is integral to the creation of any product. Whereas the craftsperson would have been both designer and manufacturer of a product, the scale and economics of industrial mass production have led to an increasing specialisation of roles within industry. One of these roles is, of course, that of professional designer.

Nowadays, professional designers will have been involved in the creation of virtually all the products that we find in our homes, communities or workplaces. Much of industry — especially those sectors of industry connected with the creation of consumer products — has come to recognise the need to produce designs that are well thought out with respect to their 'fit' to the user. This has been good news for human factors specialists. Increasingly, human factors specialists have been employed to advise designers as to how best to match a product to user needs. Indeed, the number of human factors specialists employed in industry is now at an unprecedented high. These issues were discussed in more detail earlier in the book (Jordan, Chapter 17).

21.4 USABILITY

Usually, the approach taken by human factors specialists towards fitting the product to the person has been to concentrate on the issue of usability. In layperson's terms this means ensuring that the product is easy to use. More formally the International Standards Organisation (ISO) define usability as: "... the effectiveness, efficiency and satisfaction with which specified users achieve specified goals in particular environments (ISO DIS 9241-11)."

Effectiveness refers to the extent to which a goal, or task, is achieved, *efficiency*, to the amount of effort required to accomplish a goal, and *satisfaction* to the level of comfort that the user feels when using a product and how acceptable the product is to users as a means of achieving their goals. In effect, then, a usability based approach to user-centred design is one which sees the product as a tool with which users try to accomplish particular tasks without wanting to have to expend unnecessary effort or endure any physical or mental discomfort. This definition has received wide take up as a basis for much of the human factors work carried out in industry.

Human factors as a discipline has become very adept at assuring usable design. For example, there has been a lot of work done to establish the particular properties of a design that will affect usability. Ravden and Johnson (1989), for example, link usability with the design properties such as: consistency, compatibility, feedback, visual clarity and error prevention and recovery.

Similarly, a battery of methodologies for evaluating usability have been established. The majority of these were originally developed in psychology and have been adapted specially for the evaluation of product usability. An overview of some of the most commonly used methods can be found in an earlier chapter (Popovic, Chapter 3).

21.5 USABILITY ENGINEERING AS DEHUMANISATION

Usability is an important issue and usability based approaches have undoubtedly brought huge benefits to users of products. After all, what is the point of providing users with vast arrays of functions if the design of the product makes it difficult to use them to their full advantage? Nevertheless, usability based approaches are inherently limited.

The reason why they are limited is that usability based approaches tend to look at products as *tools* with which *users* complete *tasks*. However, products are not merely tools. Products are *living-objects* with which *people* have *relationships.* Products are objects which can make people happy or angry, proud or ashamed, secure or anxious. Products can empower, infuriate, delight — they have personality (Jordan 1997).

People also have personalities. Not only do they have personalities, they have hopes, fears, dreams and aspirations. These are liable to affect the way that people respond to and interact with products. Again, this may seem obvious *prima face.* However, if a being from another planet were to try to learn about the human race via the human factors literature he/she/it would probably conclude that we were basically little more than cognitive and physical processors. It is a rarity to find published human factors studies that describe users in terms that go beyond factors such as age, gender, education or profession. The difference between 'pleasure' and the 'satisfaction' component of usability is much more than a mere semantic quibble. The human factors profession has operationalised 'satisfaction' in a manner that is limited to the avoidance of physical or cognitive discomfort. This is clearly reflected in the literature. How much work has human factors done on matching products to people's personalities, their emotional responses, or their ideals? Very little! (Although see Dandavante, Sanders and Stuart (1996) and Rijken and Mulder (1996) for rare (and good) examples).

This, then, is the problem with usability based approaches — they tend to encourage the view that users are merely cognitive and physical components of a system comprising user, system and environment. The premise of these approaches is that the product must be designed such that the cognitive and physical demands placed on the users are minimised.

It is important to recognise and emphasise that usability is a very important issue. However, there must also be an awareness that usability is only *one* of the issues that will affect the overall relationship between a person and a product. The problem with usability based approaches is that they encourage a limited view of the person using the product. This is — by implication if not by intention — dehumanising.

This seems ironic — of all the people involved in the product creation process, it is the human factors specialist, with his or her roots in behavioural science, who is the person that would be expected to have the richest understanding of users. It is no longer sufficient for the profession to think of users in such limited terms. In order to represent the user fully in the product creation process, human factors specialists must take a wider view of person-centred design and look both at product use and at those using and experiencing products in a more holistic context.

21.6 HUMAN FACTORS AND PLEASURE

The rest of this chapter will outline a 'pleasure' based approach to person-centred design. This is an approach which seems to offer the scope for human factors to broaden and extend its influence on the product creation process, to move products beyond 'usable' to the stage where they are not only usable but also enjoyable, exciting and meaningful — pleasurable.

Pleasure

What is pleasure? The Oxford English Dictionary defines it as "...the condition of consciousness or sensation induced by the enjoyment or anticipation of what is felt or viewed as good or desirable; enjoyment delight, gratification. The opposite of *pain.*"

In the context of products pleasure can be defined as:

Pleasure with products: The emotional, hedonic and practical benefits associated with products.

Note that this definition of pleasure still includes the practical 'tool-like' aspects of products. The aim is to consider emotional and hedonic issues in *combination* with practical aspects, not at their *expense*.

Hierarchy of user needs — functionality, usability and pleasure

Abraham Maslow (1970) developed a 'hierarchy of human needs'. This model views the human as a 'wanting animal' who rarely reaches a state of complete satisfaction. Indeed, if a nirvana is reached it will usually only be temporary because once one desire has been fulfilled another will soon surface to take its place. Maslow's hierarchy is illustrated in Figure 21.1. The idea is that as soon as people have fulfilled the needs lower down the hierarchy, they will then want to fulfil the needs higher up. This means that even if basic needs such as physiological needs and safety have been met, people will still meet with frustration if their higher goals are not met.

Self actualisation needs	∧
Esteem needs	∧
Belongingness and love needs	∧
Safety needs	∧
Physiological needs	∧

Figure 21.1 Maslow's hierarchy of needs.

Taking the idea of a hierarchy of needs and applying it to human factors, the following hierarchy of needs is proposed.

Pleasure	∧
Usability	∧
Functionality	∧

Figure 21.2 Hierarchy of user needs (from Jordan 1997).

Level 1 — functionality

Clearly a product will be useless if it does not contain appropriate functionality. A product cannot be usable if it does not contain the functions necessary to perform the tasks for which it is intended. If a product does not have the right functionality it will dissatisfy the user. In order to be able to fulfil user needs on this level, the human factors specialist must have an understanding of what the product will be used for and the context and environment in which it will be used.

Level 2 — usability

Once users have got used to having appropriate functionality, they will then want products that are easy to use. This more or less represents the situation at the moment in many product areas — people are used to well functioning products, now they expect usability too. Having appropriate functionality is a pre-requisite of usability, but it does not guarantee usability. To ensure usability the human factors specialist must have an understanding of some or all of the principles of usable design.

Level 3 — pleasure

Having got used to usable products, it seems inevitable that users will soon want something more. Products that offer something extra. Products that are not merely tools, but which are 'living objects' which people can relate to. Products that bring not only functional benefits but also emotional benefits. To achieve product pleasurability is the new challenge for human factors. It is a challenge that requires an understanding of people — not just as physical and cognitive processors — but as rational and emotional beings with values, tastes, hopes and fears. It is a challenge that requires an understanding of how people relate to products. What are the properties of a product that elicit particular emotional responses in a person, how does a product design convey a particular set of values? Finally, it is a challenge that requires capturing the ephemeral — devising methods and metrics for investigating and quantifying emotional responses.

The four pleasures — a new framework for human factors

So far in this chapter, a hierarchy of user needs has been given and a definition of pleasure with products has been offered. However, neither of these really gives much of a feel for what pleasure really means experientially or how it could be applied in the context of products.

Unfortunately, the human factors literature isn't helpful. This is not surprising — after all one premise on which the arguments presented in this chapter are based is that pleasure is an issue that has been largely ignored by the human factors profession. In the absence of human factors literature on the issue, the most appropriate starting point would seem to be literature from the behavioural sciences.

A useful way of classifying different types of pleasure has been espoused by anthropologist Lionel Tiger. Tiger has made a study of pleasure and has developed a framework for addressing pleasure issues (Tiger 1992). The framework models four conceptually distinct types of pleasure — physical, social, psychological and ideological. Summaries of Tiger's descriptions of each are given below. Examples are added to demonstrate how each of these components might be relevant in the context of products.

Physio-Pleasure

This is to do with the body — pleasures derived from the sensory organs. They include pleasures connected with touch, taste and smell as well as feelings of sexual and sensual pleasure. In the context of products physio-pleasure would cover, for example, tactile and olfactory properties. Tactile pleasures concern holding and touching a product during interaction. This might be relevant, for example, in the context of a telephone handset or a remote control. Olfactory pleasures concern the smell of the new product. For example, the smell inside a new car may be a factor that effects how pleasurable it is for the owner.

Socio-Pleasure

This is the enjoyment derived from the company of others. For example, having a conversation or being part of a crowd at a public event. Products can facilitate social interaction in a number of ways. For example, a coffee maker provides a service which can act as a focal point for a little social gathering — a 'coffee morning'. Part of the pleasure of hosting a coffee morning may come from the efficient provision of well-brewed coffee to the guests.

Other products may facilitate social interaction by being talking points in themselves. For example a special piece of jewellery may attract comment, as may an interesting household product, such as an unusually styled TV set. Association with other types of products may indicate belonging in a social group — Porsches for 'Yuppies', Dr. Marten's boots for skinheads. Here, the person's relationship with the product forms part of their social identity.

Psycho-Pleasure

Tiger defines this type of pleasure as that which is gained from accomplishing a task. It is the type of pleasure that traditional usability approaches are perhaps best suited to addressing. In the context of products, psycho-pleasure relates to the extent to which a product can help in accomplishing a task and make the accomplishment of that task a satisfying and pleasurable experience. For example, it might be expected that a word processor which facilitated quick and easy accomplishment of, say, formatting tasks would provide a higher level of psycho-pleasure than one with which the user was likely to make many errors.

Ideo-Pleasure

Ideo-pleasure refers to the pleasures derived from 'theoretical' entities such as books, music and art. In the context of products it would relate to, for example, the aesthetics of a product and the values that a product embodies. For example, a product made from bio-degradable materials might be seen as embodying the value of environmental responsibility. This, then, would be a potential source of ideo-pleasure to those who are particularly concerned about environmental issues. Ideo-pleasure would also cover the idea of products as artforms. For example, the video cassette player that someone has in the home, is not only a functional item, but something that the owner and others will see every time that they enter the room. The level of pleasure given by the VCR may, then, be highly dependent on how it affects its environment aesthetically.

21.7 CHALLENGES FOR HUMAN FACTORS

Having a hierarchy of user needs, a definition of pleasure with products and now a framework for looking at different types of pleasure seems a constructive basis from which human factors can move towards tackling pleasure issues. However, this basis does not of itself indicate how human factors as a profession can contribute to the creation of pleasurable products. In order to achieve this, there are at least three issues that the profession must tackle. These are understanding users and their requirements, linking product properties to emotional responses in order to fulfil these requirements, and developing methods for the investigation and quantification of pleasure.

21.8 UNDERSTANDING USERS AND THEIR REQUIREMENTS

Earlier, it was asserted that usability based approaches tend to encourage human factors specialists to consider people as processors — physical processors with attributes such as strength, height and weight, and cognitive processors with attributes such as memory, attention and expectations. Here, then, the user is often looked at as being simply a cognitive and/or physical component of a three component system — the other two components being the product and the environment. It could be argued that the traditional human factors approaches to people ignores the very things that make us human — our emotions, our values, our hopes and our fears.

In order to find a way into these issues, we need to have an understanding not only of how people use products, but also of the role that those products play in people's lives. This gives a chance to understand how the product relates to the person in a wider sense than just usability and can help the human factors specialist in gaining a wider view of the user requirements — the requirements for pleasure.

As an illustration of the pleasure benefits that products can bring to their users, three case studies are given below. These case studies were documented from interviews conducted as a 'first pass' at tackling the issues that influence product pleasurability.

Case study 1: The hairdryer user

This seventeen year old woman chose a hairdryer as her pleasurable product. A product which she described as being "perfect...the best (hairdryer) I've ever had." The reason she was so positive about the hairdryer was because it helped her to style her hair in just the way she wanted. This made her feel attractive and gave her a feeling of self confidence when she went out. She also mentioned that the hairdryer had an unusual design and was thus something that caught the attention of her friends, "...it's 'showy', I like it when people come into my room and see it."

Both of the pleasures she mentions fall in the socio-pleasure part of Tiger's framework. They are both concerned with the enhancing the image of herself that she projects to others (somebody good-looking, somebody with interesting tastes) and how she feels in the company of others (self confident).

Case study 2: The guitar player

The guitar player was a twenty-six year old man who played the electric guitar. Again, the guitar facilitated socio-pleasure. He regarded it as a "...status symbol, particularly among people who know about these things." It had also provided a talking point as it had belonged to Lloyd Cole — lead singer of Lloyd Cole and the Commotions, a Glaswegian band who had a number of hits in the 1980s including the single 'Brand New Friend'.

He also found playing the guitar an exciting activity in itself — a psycho-pleasure. Even just having the guitar near him gave him a feeling of reassurance, another psycho-pleasure.

Case study 3: The video cassette recorder (VCR) user

She was a twenty-six year old living alone who rented a VCR which she described as being a "standard video". She described the emotional benefits that she gained from the video as being a feeling of anticipation — looking forward to watching what she had recorded (a psycho-pleasure) and freedom — not having to stay home to in order to catch her favourite programmes (a socio-pleasure as she could go out with her friends).

Even in this little selection, the benefits that people mentioned went well beyond the issues that would normally be associated with usability. Ensuring that these products were effective, efficient and satisfying to use would stop well short of providing the benefits mentioned by these three people. It is clear, for example, that in case studies 1 and 2, the person/product relationship went well beyond that of user/tool. The hairdryer was an 'object d'art', the guitar was a status symbol, a talking point, even an old friend.

The first challenge is to investigate and make an inventory of the types of pleasure that products can potentially bring to their users, based on a holistic understanding of the role which the product plays in a person's life.

It should be fairly straightforward to move forward here. In the first instance it may simply be a matter of asking the appropriate questions during evaluations and requirements capture sessions. Not simply asking about functional benefits but also about the types of pleasure that a product can bring. Certainly, pleasure is not a straightforward issue. Certainly, people may not always be able or willing to articulate descriptions of the types of pleasure that they gain from a particular product. And certainly what people *do* say may often be difficult to interpret. Nevertheless, simply having an awareness of pleasure issues and asking some sensible questions will very quickly move human factors in the direction of meeting this challenge.

21.9 LINKING PRODUCT PROPERTIES TO PLEASURE BENEFITS

Having established the different types of pleasure that people can get from products, the next stage is to link those to particular product properties. For example, it might be that feelings of security with a product are linked to high levels of usability and/or high levels of product reliability. Similarly, a feeling of pride may be linked to, say, good aesthetics. Equally, particular types of displeasure may also be linked to inadequacies with respect to particular product properties: annoyance might be linked to poor technical performance while anxiety may be related to a lack of usability.

All the above suggestions are, of course, merely speculations. This is an issue that human factors must address systematically if it is to make a significant contribution to the development of pleasurable products. One way in which to approach this would be to correlate people's pleasure responses to a product to the 'goodness' of the product with respect to various properties. This approach is at the basis of a Japanese technique known as Kansei Engineering. This technique is discussed in one of the next chapters (Jordan, Chapter 23).

A less formal approach is the case study: discussing users' experience of a product and how the properties of the product are linked to their responses to it. Below are some examples, again drawn from 'first pass' pleasure with products interviews.

Case study 4: The CD player owner

This nineteen year old man had received a stereo as a Christmas present. When asked about what aspects of the product he found particularly appealing he mentioned that the CD player was "less hassle" than a tape player — he had previously owned a tape player — as the CD would play for longer than a tape without him having to turn it over. It was also possible to put the CD in repeat play mode, which again enabled him to play music for a long period without having to interact with the player at all. When asked about the emotional benefits, he said that the product gave him a sense of freedom as he was able to set it playing and then listen to the music while doing something else.

Here there was an association between one specific feature — repeat play — and one particular emotional benefit — a sense of freedom.

Case study 5: Another hairdryer owner

This nineteen year old woman had been given the hairdryer as a present 11 years ago. She found the product particularly pleasurable for two reasons. Like the first hairdryer user she said that she could use the dryer to style her hair in the way that she wanted and thus it gave her a feeling of confidence in her appearance. Unfortunately, it wasn't clear which property of the dryer had particularly contributed to making it so suitable for styling — presumably it was some combination of technical performance, functionality and usability — but this wasn't expressed clearly.

In addition, though, she also noted that the product's reliability gave her a feeling of confidence in the product itself. Here there is a link between the property of product reliability and the emotional response of confidence in the product. It might be assumed that designs which radiate a feeling of reliability are likely to trigger a feeling of confidence in the user.

Case study 6: The TV watcher

This nineteen year old woman chose her parents' television as her pleasurable product. She described this as being big and straightforward with an easy to use remote control. She said that she particularly liked the TV because of its simplicity, because of the large screen and the good picture quality, and because of the wide variety of programmes that she could choose from (her parents had a cable TV subscription). Again, though, she didn't link these qualities to any particular emotional response, other than indicating that, taken together, they led to her feeling "satisfied" with the TV.

In addition, however, she mentioned that the TV was very reliable and that this enabled her to take the TV for granted. Here, there was a link between the property of reliability and a feeling of 'security' that nothing would go wrong.

21.10 DEVELOPING METHODS AND METRICS FOR THE INVESTIGATION AND QUANTIFICATION OF PLEASURE

A major aid to the incorporation of usability issues in the design process has been the ability to quantify these issues. In the case of effectiveness and efficiency this has usually meant taking performance measures of a user with a product. These include, for example, task success and quality of output (effectiveness) and time on task and error rate (efficiency). Quantitative attitude scales, such as the System Usability Scale (Brooke

1996) and the Software Usability Measurement Inventory (Kirakowski 1996), have also been developed to measure the satisfaction (comfort and acceptability) component of usability.

Being able to quantify usability has enabled human factors specialists to set usability specifications which have then been included as part of the overall product specification. As well as giving clear usability targets to aim at, this approach has proved effective as it has given clear signals to others involved in the product creation process as to what is meant by usability in a particular context. Talking in terms of 'usability' or 'quality of use' can seem vague and 'woolly', particularly to technical colleagues, who are used to numerical specification. Quantifying the issues gives an unequivocal signal as to what is required and enables an equally firm judgement as to whether or not the criteria have been met. It follows that if human factors as a profession is to take the lead in ensuring product pleasurability, then the development of measures and tools for quantifying pleasure will be beneficial if not essential.

One approach to this would be to develop attitude scales for measuring pleasure. This requires a knowledge not only of the potentially pleasurable benefits that could be associated with product use, but also an idea of the comparative importance of each and the relationship between the benefits — which benefits are conceptually separate in the context of associations with products (i.e. which pleasure benefits are potentially, if not always, independent from each other?).

Another way to investigate pleasure with product use may be to look at potential behavioural correlates to pleasure and displeasure. For example, the frequency with which a person smiles when using a product may be seen as a simple measure of pleasure. Similarly, frowning may be seen as a simple measure of displeasure. Whilst measures such as these may be simple to take and appear to have a degree of objectivity that is lacking in questionnaire or interview based approaches, they still appear, *a priori*, to have a number of drawbacks associated with them. Firstly, there are some pleasure benefits with which smiling may not be associated — many of the ideo-pleasures for example. Secondly, there are many reasons for which a person may smile that have nothing to do with what is being experienced with the product. Perhaps they are thinking about something else, perhaps they are amused at how awful the product is!

Despite these drawbacks, facial expressions do seem a promising way forward. Facial expressions have been used as an emotional metric in psychology and there is a precedent for their use in human factors in the domain of human computer interaction (HCI). For an example of the former see Ekman and Friesen's (1978) proposed emotion coding system, while Oatley and Ramsay (1992) provide a good example in the context of HCI.

Other simple metrics of pleasure may be the frequency with which people use a particular product — where use of the product is voluntary — and hopefully, from the point of view of those who manufacture pleasurable products, purchase choice.

Of course creating pleasurable products requires more than just effective evaluation methods. It is also very important to have methods available for capturing user requirements with respect to pleasure as a starting point in the product creation process. Similarly, methods for early concept evaluation are also needed.

Human factors already has many methods that are likely to be effective here. For example, the interview, focus group and questionnaire would surely be staples for anyone addressing these issues. Nevertheless, it would be prudent to look at other potential methods for addressing these issues. One promising new method is the Private Camera Conversation, developed by de Vries, Hartevelt and Oosterholt (1996). This involves participants giving monologues to a camera in response to written questions. De Vries *et al.* claim that this method can provide rich information about users' relationships with products — participants raise issues that they may seem reticent to talk about face to face with an investigator for fear that it may appear foolish relate to a product emotionally.

Another potentially promising method is the repertory grid (Kelly 1955). Baber (1996) suggests that the approach would be useful as a means of defining users' conceptions of usability. Presumably, the same could hold true for defining concepts of pleasure with products.

21.11 CONCLUSIONS

People always have and always will seek pleasure. The artefacts and products with which we surround ourselves are potential sources of pleasure. They should be designed with a view to how they can provide pleasure to those who use and experience them. Human factors, as a profession, is in a unique position to be able to influence the creation of pleasurable designs.

Human factors has become very adept in linking design decisions with usability and, thanks to the development of a variety of methods, in the investigation of usability issues. Similar approaches can be taken with respect to product pleasurability. However, in order to be able to address pleasure issues effectively, three major challenges must be met:

- Understanding users and their requirements — understanding the emotional and hedonic benefits that a product can bring to users.
- Linking product properties to pleasure benefits — which properties of a product are associated with which benefits.
- Developing methods and metrics for the investigation and quantification of pleasure — the ability to quantify usability has been a central reason for the ever growing influence of human factors on the product creation process. A similar approach is required if colleagues in other disciplines are to be persuaded to take pleasure seriously as an issue.

Over the last two or three decades, human factors as a profession has achieved a great deal in terms of enhancing the usability of products. The profession can now move a stage further. By moving beyond usability to take a holistic pleasure based approach, the profession can facilitate the creation of useful, usable *and* pleasurable products that will delight those who use and experience them.

21.12 REFERENCES

Baber, C., 1996. Repertory grid theory and its application to product evaluation. In *Usability Evaluation in Industry*, P.W. Jordan *et al.* (Eds), Taylor & Francis, London.

Brooke, J., 1996. SUS, a quick and dirty usability scale. In *Usability Evaluation in Industry*, P.W. Jordan *et al.* (Eds), Taylor & Francis, London.

Dandavante, U., Sanders, EB-N. and Stuart, S., 1996. Emotions matter: user empathy in the product development process. In *Proceedings of the Human Factors and Ergonomics Society 40th Annual Meeting — 1996*. Santa Monica: Human Factors and Ergonomics Society, pp. 415-418.

Ekman, P. and Friesen, W.V., 1978. *Facial Action Coding System.* Consulting Psychologists Press Inc., California.

ISO DIS 9241-11, Ergonomic requirements for office work with visual display terminals (VDTs):- Part 11: *Guidance on Usability.*

Jordan, P.W., 1997. Products as personalities. In *Contemporary Ergonomics 1997*, S.A. Robertson (Ed.), Taylor & Francis, London.

Kelly, G.A., 1955. *The Psychology of Personal Constructs*. Norton, N.Y.

Kirakowski J., 1996. The software usability measurement inventory: background and usage. In *Usability Evaluation in Industry*, P.W. Jordan *et al.* (Eds), Taylor & Francis, London.

Maslow, A., 1970. *Motivation and Personality*. 2nd edn, Harper and Row, N.Y.

Oatley, K and Ramsay, J., 1992. Emotions while interacting with computers. In *Proceedings of the International Society for Research on the Emotions Annual Conference*, 18-20 August, Carnegie Mellon University, ISRE, Storrs, U.S.A.

Ravden, S.J. and Johnson, G.I., 1989. *Evaluating Usability of Human Computer Interfaces: A Practical Method*. Ellis Horwood, Chichester.

Rijken, D. and Mulder, B., 1996. Information ecologies, experience and ergonomics. In *Usability Evaluation in Industry*, P.W. Jordan *et al.* (Eds), Taylor & Francis, London.

Tiger, L., 1992. *The Pursuit of Pleasure*. Little, Brown and Company, Boston, pp. 52-60.

Vries, G. de, Hartevelt, M. and Oosterholt, R., 1996. Private camera conversation method. In *Usability Evaluation in Industry*, P.W. Jordan *et al.* (Eds), Taylor & Francis, London.

CHAPTER TWENTY-TWO

Understanding Person Product Relationships — A Design Perspective

A.J. TAYLOR, P.H. ROBERTS and M.J.D. HALL

University of Loughborough, Leicestershire LE11 3TU, UK

22.1 INTRODUCTION

Creating desirable products which people will want to buy, use, and live with over a substantial period of time is a complex activity requiring input from a range of diverse disciplines. The days are gone when a consumer product could be entirely created by one single designer, and successful new product development (NPD) is increasingly dependent on cross-disciplinary communication and understanding. The early, definitive stages of design involve designers, ergonomists, marketing people and others in a complex combination of activities in which the viewpoints of the various participants inevitably differ. To make progress towards a synergistic integration of the design activity there needs to be an ongoing dialogue between the disciplines. This chapter is intended to contribute to that dialogue.

Much of the material in this chapter is based on research at Loughborough University in the last two years to investigate semantic characteristics of electronic products in terms of added value and perceived quality. This work set out to contribute to our insight into the ways in which electronic consumer products convey meanings and values to users and was based on the premise that, with the continuing reduction in size and increasing power of electronic systems, the external form of consumer products no longer has to be defined or heavily constrained by functional components. The human context, therefore, has become the driving force for design.

A product must not only function satisfactorily but must embody symbolic qualities appropriate to its intended user group and environment which express something more than overt functionality. The ways in which product qualities and characteristics are transmitted to the user are complex, and the ways in which these qualities are interpreted by different users are influenced by numerous factors such as age, culture, education and context of use.

Persuasive arguments have been put forward in recent years by many writers, from both academia and industry, for product semantics to be considered as a central issue in design. A concise summary of the main arguments is provided by Langrish (1996). Krippendorff (1990) sums up the aims of product semantics in a proposed new paradigm which is user-enabling and based on a philosophy of designer/user cooperation rather than the 'old' paradigm of designer as an authority and in which the user is expected to conform to some behavioural stereotype. Products should communicate their purpose and means of use in ways which allow intuitive use and positively encourage exploration of the product's capabilities. The new paradigm is a desirable goal, but in practice is easier said than done. The study of product semantics might now be said to be at a stage where the 'what' and the 'why' are becoming established; the problem now is 'how'.

The literature search, initially based on product semantics, aimed at identifying relevant theories, underlying principles and ideas from a variety of fields. This approach led to a broadening of the information base, but became highly divergent in that almost everything encountered was found to have some potential relevance to human perception of products, and it appeared to be impossible to establish any boundaries, except on some arbitrary and limiting basis. This in itself is instructive since it highlights the eclectic and fundamental nature of this field of study.

Our research made use of a combination of quantitative and qualitative methods, the later work taking a bottom-up approach focussing on personal stereos and telephones to help in identifying key issues.

Most of the observations discussed here are the results of interviews, both with users and with practitioners in design, ergonomics and marketing.

22.2 SOME FINDINGS OF THE RESEARCH

The investigation has confirmed the importance of emotive and psychosocial responses to designs, and has highlighted the need for a basis on which design teams can quickly acquire an improved understanding of these issues. This is by no means easy, with difficulties arising both from the inherently complex nature of designing (design is done by people for people) and the broad diversity of factors which can influence the person product relationship. The hedonic values associated with products are central to industrial design (sometimes termed the 'X factor') but inevitably present a somewhat elusive area for research, being subject to behavioural, sociocultural and psychosocial influences as well as task-oriented responses to a product.

Consideration of the difficulties encountered, together with literature by Jordan (1997) and others, lent weight to the notion of a three-stage categorisation of product characteristics or qualities: functionality, usability, and the hedonic effects. These three interact and are not entirely separable but may be regarded as distinguishable, thereby providing a basis for further understanding, discussion and investigation. The influencing factors for the third area of pleasurable or experiential responses are, not surprisingly, difficult to isolate since they are clearly affected by functionality and usability. A holistic view is therefore needed and, considering the breadth and complexity of the subject area, it is arguable that the only appropriate vehicle to achieve this will be the product itself.

As part of a product based approach to the investigation, the perceived quality of six personal stereos was examined by using the method of paired comparisons with six interviewees. A temporal breakdown based on the three stages of initial impact, familiarisation and regular use was adopted, the first two being considered in this trial. The initial impact stage was split into two: a purely visual examination followed by physical handling.

It was found that initial perceptions of quality can change significantly once the products are handled. For example, the tactile and auditory qualities associated with opening and closing the player and loading a cassette work together to create a response which, in the case of the best and worst products in the range, is very emotive — "that's really nice" or "that's awful". Attitudes to the product can change after handling, influencing the previous visual perceptions — in other words, people change their minds.

These first tests were followed by a series of interviews using eight office telephones and four mobile phones. This confirmed that emotive qualities are a powerful influence, even for products as utilitarian as an office telephone. People readily use visual and imaginative associations in their perceptions of products such as telephones and there is inevitably overlap between the functional values and hedonic values. The original intention of the office telephone interviews had been to first of all obtain responses to the complete product in order to identify particular product characteristics which evoked a strong response, and then to find out whether it is possible to investigate these as isolated features. This in fact was found to be very difficult. Isolating design features for examination in most cases destroys the ecological validity of the tests to be carried out because of the importance of the totality of the product and the interactive nature of its various characteristics. Relating specific properties of a product to particular emotions or responses is not at all straightforward.

Jordan (1997) refers to the challenge of linking product properties to emotional responses and proposes Tiger's (1992) framework of four distinct types of pleasure (physiological, sociological, psychological and ideological) as a starting point (see Jordan, Chapter 21 for a description of each of the pleasure types). Relating this approach to the results of the telephone interviews was found to be helpful in bringing to light a wide variety of issues, of which the following are a few examples.

Physiological pleasure

Physiological pleasure, in terms of responses to the tactile qualities and operation of the keys, constituted a major influence; out of eight factors listed keypad design was found to have the highest influence on perceived quality and product desirability. A positive key action without excessive lateral play was important, as was the material of the keys. Auditory qualities can tell us something of the quality of build; for example, any creaking of the plastic body of the handset had a strong negative effect.

Sociological pleasure

Connotations of sociological pleasure are inherent in the telephone as an instrument of social interaction. Many people commented that they do not think consciously about the phone itself provided it allows them to easily make a call — they are more concerned with the conversation than the instrument. We could say that many products today are taken for granted but missed when they are not there. Having a telephone to hand is part of a way of life in which we expect to be able to communicate readily over any distance.

Mobile phones have more powerful emotive characteristics than office telephones — they are personal items and represent an enhanced communication capability. They have, to date, been regarded as implying a particular lifestyle (someone who needs to be contactable in a variety of circumstances) and therefore conferring a certain level of importance on the user. Responses to products are closely linked with a person's sense of identity and self-image — "what does owning or using this product say about me?" This is clearly the case with mobile phones but the same consideration applies generally to some degree.

Psychological pleasure

Psychological pleasure in terms of usability is clearly important for telephones. We found that perceived complexity was considered daunting, but an LCD screen was found to provide reassurance as people regarded it as, potentially, providing a source of helpful information. In considering usability people with long fingernails sometimes appeared to be overlooked by designers. Mobile phones which could easily be used one-handed by dialling with the thumb were preferred by all interviewees. Less than eight hours charge in the battery detracts from the pleasure of ownership by not allowing a full day's use.

Ideological pleasure

Ideological pleasure, the most abstract of the four, was found to be important even for a product as utilitarian as an office telephone in that the aesthetic qualities elicited clear responses from all twenty interviewees. It was noticeable that all tended to use associative imagery to imagine each of the eight telephones in a particular office context, and being used by imaginary stereotypical people. The perceived age of the product was very influential in terms of being modern (hence desirable) or old fashioned, and, for no conscious reason, older telephones were automatically associated with older users. One colour, a cream hue, raised unprompted concerns of hygiene standards, even though the item itself was clean and had a gloss surface.

Overall the four-pleasures approach appears to provide a valuable basis for further investigation. Some overlap was evident between the four categories; tactile feel and feedback of the keys affected physio-pleasure and psycho-pleasure aspects, and similarly socio-pleasure and ideo-pleasure were often seen to interact.

22.3 THE NEED TO UNDERSTAND INDUSTRIAL DESIGN

The main aim of industrial design is to add value to a functional device (in other words to progress from a device to a product) and therefore it could be said that to attempt to understand emotive responses to products is to attempt to understand the very basis of industrial design itself. This, of course, is extremely ambitious and might be said to be attempting the impossible. However, the ultimate aim of research need not be to arrive at prescriptive design methods, even if that were possible, but to help develop assistive methods or techniques which help design teams do more effectively what they already do. Design in reality makes use of strategies more than methods, and strategic thinking depends on understanding.

Acquiring an in-depth understanding of human cognitive and emotional responses to products will need to draw on areas such as cognitive and social psychology, art, cultural influences and other fields of study, some of which are in the formative stages themselves. Add to this the extraordinary adaptability of people as users, resulting in products themselves having the effect of altering a person's perceptions and attitudes, and the scale of the problem becomes apparent. It is in fact remarkable that in practice products are being designed, often very effectively, on a daily basis.

Gaining knowledge and understanding is an incremental process, and there is already a large body of knowledge. One valuable and readable account of how human perception relates to product design is provided by Crozier (1994). This knowledge, so far as design is concerned, is not an end in itself but needs to be encapsulated and embodied in product entities. Therefore, although we need to understand more about people — their perceptions, attitudes and aspirations, their responses to products, the roles the products play in their lives — we also need to understand design — how this knowledge and understanding may be assimilated and used by multi-disciplinary teams and ultimately embodied in the artefacts which are produced. This second step must not be overlooked.

Product design can be viewed from different perspectives, and differences in approach inevitably exist between designers, ergonomists, marketers and production engineers. Most disciplines have their established analytical methods and techniques since they are analytically based. Industrial design is more synthetic and has less formal methodology, but must, by its very nature, encompass many issues crucial to an understanding of person product interaction. Designers work with the links between design characteristics and emotive response as part of their everyday work and therefore, although not easy to articulate, they must already have some of the answers. To take some fairly obvious examples, horizontal lines or features imply stability and may have a calming influence, diagonals are dynamic and can be disturbing. Colours, and subtle shade differences, are used by designers to evoke particular feelings and attitudes. Weight can imply high build quality and provide reassurance. These factors are, of course, context-dependent and by no means absolute, but context is in any case a fundamental design consideration. Theories for achieving pleasing visual effects in design have existed for a long time, having been applied notably by the ancient Greeks. The golden section and the good Gestalt are enduring concepts today. It seems likely that aspects of designers' approaches and philosophies could form a valuable input to research into hedonic responses to modern products.

22.4 CASE STUDY — THE ORION TELEPHONE

The following case study was carried out to explore aspects of the design activity itself, and examined how a major telephone manufacturer, Nortel, and an industrial design consultancy, Smallfry, worked together to produce a successful office telephone design.

The starting point for the Orion design was the Aries telephone designed by Nortel in Canada for the North American market. This was to be redesigned for the European market. The internal components and the handset were to be retained but the form was to be completely different.

The Aries and Orion products are shown in Figures 22.1 and 22.2.

Figure 22.1 Aries telephone.

Figure 22.2 Orion telephone.

The design process

Following the initial brief twenty concept renderings illustrating a wide range of ideas were provided by Smallfry for evaluation by a group of twelve people at Nortel comprising product line managers, designers, project managers, and sales and marketing people. Features or ideas which the group felt should be taken forward were identified from several of the designs. The open discussion allowed a wide range of issues to be considered from apparent stability of the form to issues of tracker ball versus cursor keys. It was found to be difficult to make design choices with a group of this size, but certain features emerged on which a consensus could be reached. These included the positioning of the LCD screen and the elliptical cross-section of the body which is a distinctive feature of the final design.

Smallfry then amalgamated these desirable features within a new design proposal and from this a final design was achieved by a process of refinement. This stage took around four weeks. Design for production was then carried out by Nortel, with continued dialogue between Nortel and Smallfry. A telephone typically takes two to three months for detail design once the overall form has been defined.

Design evolution

A significant change between the Aries and Orion was the keypad layout. The designers considered it important that the key spacing should be increased and the proportions of the key shape altered. Also the keypad is visually separated from the other keys by a recess. The reasons for this were largely aesthetic, the relationships between the keypad layout, the other visual features and the overall form of the phone body being of vital importance. This is an illustration of the important point that, in the perception of products, visual characteristics have to work together to create the required Gestalt and subtle changes can have a major impact on the overall effect. This change was very significant to Nortel in terms of both cost and product identity, but after serious consideration it was agreed that the proposed dimensional changes were essential to the quality of the new design and a commitment was made to produce the new keypad components. The same tactile qualities, which had been successful previously, were retained by using the same materials and operating parameters. Corporate identity was maintained by the use of existing colours, fonts and materials.

The hook switch is a detail design aspect which required careful development work, particularly in terms of spring strength, to ensure correct functioning in both desk and wall mounted modes. Achieving optimum functioning of details such as this can be time-consuming, but is very important in terms of customer satisfaction since there is no doubt that poor or uncertain handset location is the type of problem which could cause considerable annoyance to users and badly affect perceptions of the product.

Surface texture is an important consideration, and is closely linked with colour. In fact the trials at Loughborough indicated that colour and texture tend to be perceived as a single characteristic. A textured surface was used on the Aries and Orion. When design for manufacture at Nortel necessitated some very minor modifications to the form the industrial designers were always consulted because even slight changes may affect the reflections and hence have a significant effect on visual perceptions.

Design evaluation

Fifteen pre-prototypes were produced to allow testing in the factory and also several working prototypes supplied to representative customers for initial field trials. Following this two thousand sets were supplied to selected customers for field trials before the model was released for general sale. Modifications necessary following these trials were of a very minor nature.

The interviews at Loughborough using eight different office phones showed the Orion to be the most desirable of the products, with six of the twenty people interviewed making it their first choice.

22.5 DESIGN PHILOSOPHY

The following observations were collected from the designers interviewed during the study:

On product development

"Product development is always about speed".
"Communication with the client is important, and particularly with Marketing people".
Designers must be able to speak to people at all levels and cross-functional team communication is vital.
Computer simulation is very useful in communicating ideas. For evaluating final design proposals 3-dimensional physical models which can be handled are essential. It is important to appreciate that 3-dimensional design is not 2-dimensional design expanded to three views. To think in 3-dimensional terms there is a need to sculpt.

User perceptions

Users' visual perceptions of the quality of a product are based entirely on surface information. Therefore surface qualities are of paramount importance.

It is difficult to differentiate between customer's needs, musts and wants, and it is difficult to find out about, or measure, customer's perceptions without influencing them, particularly for new products.

Factors influencing commercial success

Industrial designers are not artists in the sense of fine art (though industrial design may be an art) — they design things for commercial benefit to companies. The perceived value of a product may be improved by 35% to 40% in some cases, but with an increase in production costs of perhaps 5%.

The user and customer may be different parties. It is vital to understand the design context and who the customers are — good design is appropriate, and satisfying the customers is the single most important factor.

In companies strong links between the marketing and design functions are important, but in practice this does not always happen. The most professional and successful companies can be seen to link marketing and design. However, many other factors have significant influences on a design, such as short-term finance pressures, internal company politics and legislation.

Techniques and strategies

Smallfry have methods and techniques for effective design which have been developed over a number of years and, not surprisingly, the details are highly confidential. In principle they involve bringing together a broad range of considerations covering costs and manufacturability together with user based issues of perceived quality and customer benefits in ways which allow maximisation of the profitability of the final design. Underlying much of the design approach is the notion that people do not buy features, they buy benefits.

22.6 THE DESIGN PERSPECTIVE

A design perspective may be considered as being distinct from those of ergonomics, marketing or production engineering, and we believe that clarifying these viewpoints may contribute towards a holistic understanding of the broad, complex and eclectic nature of product development.

We have therefore attempted to identify some of the essential elements or characteristics of the design activity as a basis for discussion and comparison with other fields or viewpoints. The following observations emerged. Some of these may appear obvious but we believe that there is value in making these explicit.

- Designers work in the future tense — they need to see things as they might be rather than as they are. Designing is generative rather than evaluative (although evaluation is an integral part of the creation process). In contrast, most research methods are based on current or past events or data. For example, as Lorenz (1990) illustrates, too much reliance on market research using focus groups or clinics can lead to staid products or, as Peters (1995) puts it, "the customer is a rear-view mirror, not a guide to the future". Designing has a strongly proactive element, not simply responding to the needs and wishes of the users and buyers.

- Products are perceived by people as entities in which the various elements of the design are interactive (the Gestalt). In investigating responses to product designs, therefore, isolation of particular properties or features for examination is not always possible or valid. Colour and surface texture is a common example — we tend to perceive something as, say, matt black or shiny red.

- All designing is concerned with the creation of *entities* through an essentially evolutionary process. This entity must embody the functional and human interface characteristics relevant to particular groups and particular contexts. [In teaching product design Mike Hall would sometimes say to a student "this design has potential but you haven't quite got it under control" — not a definitive statement by any means but the fact that it was meaningful to the student is indicative of the holistic and emergent nature of design.]

- The aesthetic effect of a product may be altered significantly by subtle changes, especially to form or colour. Form is perceived largely by reflections which may be altered by small changes to a 3-dimensional form.
    Form (in this sense) is not the same thing as configuration. Two products with virtually identical configurations can invoke differing aesthetic responses, just as the difference between a beautiful face and a plain face may be almost immeasurable in terms of any quantifiable method.

- In practice the design process is usually characterised by very severe time constraints (as illustrated by the Orion case study).

- Design information needs to be available in an appropriately concise and readily assimilated format.

- The act of designing makes use of much non-verbal information, and many design phenomena are, themselves, not susceptible to linguistic representation. Designers work in largely non-verbal, non-textual ways. Therefore their ways of understanding and the reasoning behind design decisions are not always easy to articulate.

- Design is done by people for people. Since humans are extremely complex this means that design is inherently complex.

- Design is always context-dependent, both environmental and human.
    Perceptions and attitudes to products change over time, influenced by personal factors as well as by new products entering the marketplace. The design context is therefore dynamic and we are aiming at moving goals.

22.7 IMPLICATIONS OF THE DESIGN PERSPECTIVE

The possibility of directly relating product 'properties' to emotive responses is problematic because of the subtlety, and hence indefinability, of some of these properties, and also because of the interactive, holistic nature of the product entity. (Although tools such as SEQUAM (Bonapace, Chapter 24) and Kansei Engineering (Jordan, Chapter 23) represent attempts to do this).

The severe time constraints inherent in commercial design often preclude the possibility of extended trials or research on particular aspects of design, as might be carried out by ergonomists or market researchers. Knowledge in practice tends to be built up over a number of products and to an extent transferred from one to another.

To ask why some products succeed and some fail in the marketplace is clearly a pertinent question. However, from a design viewpoint a more pertinent question is "How can we make the next product succeed?". These two questions, though linked, are not exactly the same. We cannot say that thoroughly examining users' responses to an existing product will allow us to predict with any certainty their response to a new design, even if it is very similar. There has been an implicit, and perhaps not unreasonable, assumption in the literature and research (including this project initially) that obtaining a more comprehensive understanding of people's perceptions and responses to existing products will enable us to design new products more successfully. However the processes involved in extrapolating knowledge and understanding from existing products to the design of a new product (even if only incremental changes are being considered) are not necessarily straightforward. History indicates that knowledge and experience from previous designs is no guarantee of success in designing a new generation of products. As mentioned before this second, predictive, stage should not be overlooked.

To understand human emotive and hedonic responses to products we need to draw on concepts and research techniques from a variety of fields including (but not limited to) design, human factors, marketing, cognitive psychology, social psychology, anthropology and art. Clearly any participant in the NPD process cannot attempt to become an expert in these other fields, and for effective communication it will therefore be necessary to:

1. distil the relevant information, understanding and techniques and make the results accessible in a concise 'design information' format, and;
2. provide a basis of understanding by which this knowledge can be utilised in developing new products.

Because of the breadth and complexity of the subject we need to move towards a framework of understanding.

22.8 CONCLUSIONS

Our research has shown that once a knowledge and understanding of the customers has been gained, the embodiment of this understanding in new products is not as straightforward as might be assumed and is still largely dependent on the 'black art' of the industrial designer. We need to develop our understanding not only of customers and users but also of the design activity if we are to better understand the complex and fundamental issues of person product relationships. Considerable knowledge already exists. The challenge, as well as developing this knowledge, is to develop the means by which it can be effectively embodied by design teams in new artefacts and products.

Prescriptive design methods may not be possible (or desirable), we should aim to provide design teams with means to enhance their own understanding of the issues. The product entity provides a focus and boundaries for understanding the human issues; without this the problems become too general and large to deal with.

The discourse between the various disciplines relevant to NPD is important and should be developed, and in particular the synergy between designing, ergonomics and marketing in the conceptual design stages.

22.9 REFERENCES

Langrish, J., 1996. Product semantics and ease of use. In *Proceedings of the 8th Internatonal Forum on Design Management*, November 1996, Barcelona.

Krippendorff, K., 1990. Product semantics: a triangulation and four design theories. In *Proceedings of Product Semantics 1989*, S. Vakeva (Ed.), University of Industrial Arts (UIAH), Helsinki.

Jordan, P.W., 1997. A vision for the future of human factors. In *Proceedings of the HFES Europe Chapter Annual Meeting1996*, K. Brookhuis *et al.* (Eds), University of Groningen Centre for Environmental and Traffic Psychology, pp. 179-194.

Tiger, L., 1992. *The Pursuit of Pleasure*. Little, Brown and Company, Boston, 1992.

Crozier, R., 1994. *Manufactured Pleasures — Psychological Responses to Design*. Manchester University Press, Manchester.

Lorenz, C., 1990. *The Design Dimension — The New Competitive Weapon for Product Strategy and Global Marketing*. Basil Blackwell Inc., Cambridge, Mass.

Peters, T., 1995. Pundit of passion. *Design — The Journal of the Design Council*, Summer 1995, p. 19.

22.10 ACKNOWLEDGEMENTS

EPSRC
Steve May-Russell, SmallFry Ltd
Nortel Ltd, Cwmcarn, S. Wales: Alan Zimmerman and Roger Snook
Sam Read
Peter Simcoe

CHAPTER TWENTY-THREE

Kansei Engineering and Design

PATRICK W. JORDAN

Philips Corporate Design, Building W, Damsterdiep 267,

P.O. Box 225, 9700 AE Groningen, The Netherlands

23.1 PLEASURE WITH PRODUCTS

In a previous chapter (Jordan, Chapter 21), it was argued that human factors should look not only at, but also beyond usability. It was suggested that the profession should take 'pleasure-based' approaches to user-centred design. Pleasure with products was defined as: *the emotional, hedonic and practical benefits associated with product use*. In particular, it was argued that pleasure-based approaches encourage designers to take a holistic view of users, while usability-based approaches may encourage designers to 'dehumanise' users — to think of them as merely cognitive and physical processors.

23.2 LINKING USER RESPONSES TO PRODUCT PROPERTIES

It was also argued that if human factors is to support pleasure-based approaches to design, the discipline faces the challenge of making links between product properties and users' emotional reactions to products. This chapter is an introduction to Kansei Engineering — a Japanese technique — developed for the purpose of making such associations. This technique has received wide acceptance within a variety of design contexts. This chapter gives a brief overview of the technique, and reports two case studies illustrating its application. The advantages and disadvantages of the technique are discussed. A number of other approaches to linking product properties to user reactions are then summarised, followed by a discussion of the comparative merits of each.

23.3 KANSEI ENGINEERING

Kansei Engineering (Nagamachi 1995) is an empirical technique aimed at linking the design characteristics of a product to users' responses to that product.

Kansei Engineering can work in two ways — known as the two directions of 'flow'. One direction of flow is termed 'from design to diagnosis'. This involves manipulating individual aspects of a product's design in order to test the effect of the alteration on users' overall response to the product. This technique has been used to assist in the design of a diverse range of products. Nagamachi (1997) describes examples that range from automobiles through camcorders to clothing.

To demonstrate how the technique works, a case study — reported by Ishihara *et al.* (1997) — is summarised below demonstrating how Kansei Engineering was applied to the design of cans for coffee powder.

The other direction of flow is from context to design. This involves looking at the scenarios and contexts in which the product is used and then drawing conclusions about the implications of this for the design. This is not unlike the human factors approaches commonly taken in the west. The difference is one of scope — Kansei approaches look at issues that are wider than usability.

In the next section of the chapter, two case studies are reported, demonstrating the application of Kansei Engineering in each of these directions of flow.

23.4 EXAMPLES OF THE APPLICATION OF KANSEI ENGINEERING

Kansei flow 1: from design to diagnosis

This technique has been used to assist in the design of a diverse range of products (Nagamachi 1997). To illustrate this type of approach, an example is described below demonstrating how Kansei Engineering was applied to the design of cans for coffee powder.

In this study — reported by Ishihara *et al.* (1997) — 72 alternative designs of coffee can were presented to a panel of ten subjects. Each member of the panel was asked to rate each of the designs according to how they fitted with a series of descriptor adjectives. These adjectives are known as 'elements'. There were 86 of these elements. Panellists rated each of the 72 designs according to the 86 elements by marking 5 point Lickert Scales to indicate the degree to which they felt each of the designs exhibited each of the elements. This meant that each participant made a total of 6,192 responses on Lickert Scales! Examples of elements used in this case were: showy, calm, masculine, feminine, soft, individual, high-grade, sweet, milky etc.

A cluster analysis was then carried out in order to determine how panellists tended to group the designs according to their elements. A number of clusters emerged which Ishihara *et al.* (1997) were able to identify as being linked to particular design features. For example one cluster of cans emerged that were regarded by the panellists as being 'milky', 'soft', and 'sweet' — this cluster of cans was characterised by the use of beige colouring for the majority of their surface. Another cluster was seen as being 'masculine', 'adult' and 'strong' — these elements were all associated with having a large logo on the cans. A third example of a cluster was one that was seen as being 'unique', 'sporty' and 'individual' — this was related to the use of blue and white colouring in the designs.

Kansei flow 2: from context to design

The other 'direction' in which Kansei Engineering can be applied is to take knowledge about the users' environment and the context of product use and then to apply this to the product design. These approaches can often mirror the traditional human factors field study. Indeed sometimes the studies will involve little more than looking at the ergonomics of product use in a particular context. Often, however, the emotional aspects of product use will also be considered. Nagamachi (1997) reports on this type of approach in the context of the design of three different products — a refrigerator, a camcorder and a brassiere.

The refrigerator project, carried out for Sharp, represented the first successful application of Kansei Engineering to product design (Nagamachi 1995). In essence this was little or nothing more than a field-based approach to user-centred design with traditional ergonomic concerns at the centre. Nagamachi and his colleagues visited the homes of a number of refrigerator users in order to investigate the issues that arose in use. These mainly involved physical issues such as users having to bend awkwardly in order to reach particular items in the fridge. The recommendation, basically, was to alter the positions of the various components of the fridge. Nothing new, then, from traditional usability based approaches.

Similarly, the approach to camcorder design was simply based on a (rich) understanding of the context in which people tended to use the product. From looking at marketing data Nagamachi discovered that the biggest single user group for camcorders was families who had just had babies. This meant, for example, that users had to get down on their hands and knees in order to film a baby crawling. To solve this problem Nagamachi recommended a lens that could be swivelled through 360 degrees. Another, aspect of use that Nagamachi noticed was that users would often enjoy showing others what they had just filmed. In order to do this, the users would ask their friends to look through the viewfinder and would then play back what they had just filmed. To improve on this rather inelegant solution Nagamachi proposed adding a large, bright monitor to the back of the camcorder. This enabled others to view the playback without having to look through the viewfinder.

The approach to the brassiere design, on the other hand, did go beyond traditional usability approaches. The factors included here mirrored those included in the coffee can design. However, while in the coffee can study participants were presented with a number of concepts and asked to rate them with respect to pre-identified 'elements', the participants in the brassiere study were asked to identify the elements themselves. Five hundred women were asked about the elements that were important in terms of how wearing the brassiere should make them feel. 'Beautiful' and 'graceful' emerged as the two most important elements and subsequent design concepts were evaluated according to how they made their wearers feel with respect to these two dimensions.

23.5 SIMILAR APPROACHES

Aside from Kansei Engineering, a number of other techniques and approaches have been developed for investigating the link between product properties and user responses. Some examples are given below.

Sensorial Quality Assessment (SEQUAM) (Bandini-Buti, Bonapace and Tarzia 1997)

This technique involves analysing the link between the physical properties of a product and users' responses to tactile contact with the product. Users are presented with models of product components (e.g. car door handles) that are mocked up in combinations of materials and finishes exhibiting a range of tactile qualities. Users are then asked to handle these mock-ups and to comment on their sensorial qualities. SEQUAM has been applied in the automotive sector by Fiat.

Case studies (Macdonald 1998)

Macdonald has analysed a series of products and has suggested links between particular elements of their design and particular pleasures of use. The table below summarises the outcomes of a study in which the benefits associated with a number of products were analysed within the context of the 'four pleasure framework' (for details of the study see Jordan and Macdonald 1998). This framework identifies four separate types of pleasure with products:

1. physio-p   —   pleasures to do with the body and the senses.
2. socio-p   —   pleasures to do with relationships (inter-personal and social).
3. psycho-p   —   pleasures to do with the mind.
4. ideo-p   —   pleasures to do with values.

Table 23.1 Four pleasure analysis of the benefits associated with a sample of products from Macdonald's case studies (from Jordan and Macdonald 1998).

PRODUCT	PHYSIO-P	SOCIO-P	PSYCHO-P	IDEO-P
Karrimor's Rucsac Buckle	• positive click on closing			
Global Knives	• comfortable to hold • weight balance		• reassuringly hygienic	
NovoPen™ (for the injection of insulin)	• pleasant to hold	• plays down negative associations (from others)	• easy and reassuring to use	• plays down negative associations (from user)
Samsonite Epsilon Suitcase	• comfortable, lifting, tilting and trailing		• responsive movement — suitcase 'obeys' user	
Mazda Car Exhaust		• status symbol		• positive, youthful self-image

Semi-structured interviews (Jordan and Servaes 1995)

Jordan and Servaes carried out a series of interviews asking users about their most pleasurable products. They recorded users' reported reactions to pleasurable products and the product properties associated with these reactions. Reactions to pleasurable products included, for example, feelings of security and comfort, pride, excitement, freedom, and nostalgia. Examples of associated properties included features and functionality, usability, aesthetics, performance, reliability, and associated status.

23.6 DISCUSSION

Kansei Engineering represents a thorough, formal approach to the linkage of product properties with user responses. Like SEQUAM it relies on statistical analyses to make these links. Such techniques appear to offer the most reliable and valid means of making product property/user response links. However, when many property dimensions are involved, as in the coffee can example, such methods can become unwieldy and time consuming. Designing 72 concepts and asking respondents to mark over 6,000 scales each, simply in order to give input to the design of a coffee can, seems excessive!

Another possible criticism of Kansei Engineering is that, in analysing the effect of individual design elements, it is implicitly based on the assumption that a design is the sum of its parts. The merit of such an assumption is a matter for debate. It could be, for example, that a 'Gestalt' model is more appropriate. Gestalt theories assert that entities, such as designs, must be considered holistically — that is to suggest that the overall response to a design could amount to either more or less than the sum of reactions to each of its constituent parts.

From the Gestalt point of view more qualitative approaches — such as case studies and interviews — may be more appropriate, as they may be better suited to looking at the product as a whole. However, it might be argued that, unless design elements are separated, it is difficult to give designers any meaningful advice as to how to create a product that will elicit particular responses.

23.7 CONCLUSION

Although it may sometimes be unwieldy to apply, Kansei Engineering is arguably the most reliable and valid technique for linking product properties to user responses. The technique has a proven track record in applications covering a wide range of products and has proved an effective approach to creating designs that will delight the user.

23.8 REFERENCES

Bandini-Buti, L., Bonapace, L. and Tarzia, A., 1997. Sensorial Quality Assessment: a method to incorporate perceived user sensations in product design. Applications in the field of automobiles. In *Proceedings of IEA 1997*, Finnish Institute of Occupational Health, Helsinki, pp. 186-189.

Ishihara, S., Ishihara, K., Tsuchiya, T., Nagamachi, M. and Matsubara, Y., 1997. Neural networks approach to Kansei analysis on canned coffee design. In *Proceedings of IEA 1997*, Finnish Institute of Occupational Health, Helsinki, pp. 211-213.

Jordan, P.W. and Mac Donald, A.S., 1998. Pleasure and product semantics. In *Contemporary Ergonomics*, M.A. Hanson (Ed.), Taylor & Francis, London.

Jordan, P.W. and Servaes, M., 1995. Pleasure in product use: beyond usability. In *Contemporary Ergonomics*, S.A. Robertson (Ed.), Taylor & Francis, London, pp. 341-346.

Macdonald, A.S., 1998. Developing a qualitative sense. In *Human Factors in Consumer Product Design and Evaluation*, N. Stanton (Ed.), Taylor & Francis, London, pp. 175-191.

Nagamachi, M., 1995. *The Story of Kansei Engineering*. Kaibundo Publishing, Tokyo.

Nagamachi, M., 1997. Requirement identification of consumers' needs in product design. In *Proceedings of IEA 1997*, Finnish Institute of Occupational Health, Helsinki, pp. 231-233.

CHAPTER TWENTY-FOUR

The Ergonomics of Pleasure

LINA BONAPACE

Ergosolutions, Via Slataper 21, 20125, Milan, Italy

24.1 SENSORY PERCEPTION

People perceive the world through their senses, mainly by sight, hearing, touch and the awareness of our body in space (kinaesthetic sense), and then by odour and taste. The world can be considered as a supplier of stimuli that reach people through sensory channels and that trigger the stimulus and response mechanisms: input, perception, process (cognition), action and output.

The *input* is the stimuli that the environment sends and that can be gauged, measured and reproduced, while *perception* is how much the individual actually takes in after the signal has passed through the person's psycho-sensory filters and the meaning the person attributes to it. For example, if a person were in a classroom attending a lesson that is particularly interesting, the auditory attention can be so great that external noises are filtered because the mind of the listener is intent on grasping what is being said. If, though, the person is tired or the lesson is boring, any type of information that comes in from the outside is grasped and superimposed, mixing and interfering with what is being said in the classroom.

	EXAMPLE OF STIMULUS RESPONSE
INPUT	THE FRONT DOORBELL RINGS
PERCEPTION	I HEAR IT WITH AN AUDITORY ORGAN (if I am not deaf or distracted by other things) I UNDERSTAND THAT IT IS A BELL (comparison with formal models in the memory)
PROCESS	I DECIDE TO OPEN (if I see fit)
ACTION	I OPEN THE DOOR (if I am not prevented from doing so)
OUTPUT	THE DOOR OPENS

Figure 24.1 Example of stimulus response.

This surely depends partly on the characteristics of the listener, but also on the quality of the input. If a loud or intense sound like a siren is set off outside the window, it would be noticed by almost anyone, because the input is so important that it overcomes all the filters that may have been activated by the person's attention. The complex cognitive mechanisms of short- and long-term memory then determine the *process*, *action* and consequent *output* phases. Figure 24.1 illustrates an example of a stimulus-response process.

None of this helps to define the quality of the stimuli. People can in fact activate perception with stimuli that are effective but unpleasant, or effective and pleasant. Studies of a strictly ergonomic nature, that is, studies which concern physical and psychological well-being (posture, strain, physical and mental strain) are limited in the sense that they do not allow one to gauge the sensations felt by the individual when facing an object, a situation or simply his/her being in an environment.

Recently, however, ergonomics as a discipline has started to concern itself with aspects of products that are not exclusively connected to greater or lesser ease of use, but more attributable to its sensory aspect (see also Jordan, Chapters 21 and 23; Taylor, Chapter 22). In other words, given the same 'ergonomic' efficiency, a subject will tend to further express pleasure for the various solutions, depending on the sensation that the object transmits to by way of its sensory qualities. It seems clear then, that alongside comfort, the sensations of pleasure transmitted by the object to the individual is an important aspect in the search for quality.

If the aim were to gauge the quality of a handle, aspects concerning its use (how it is grasped, ease with which its functional direction can be assertained etc) and concerning its safety (if there are sharp edges, corners etc) would have to be assessed. However, if the handle were only to be gauged by these aspects it might be assumed that many are indistinguishable from others because they produce identical effects on the people that use them. In reality though, no one would say that the sensations from grasping a wooden handle are the same as those from grasping one in brass or ceramics. For example, in the instance of case handles an evaluation should rate their use characteristics — they have to be big enough and well shaped so as not to cut into the hand — but to gain a fuller understanding of the users' experience it is also necessary to gauge the sensation the material transmits during the contact with the fingers. The leather handle will give a sensation that is different to that of a metal product even if the shape is the same. But even coarse leather will give a sensation of touch that is quite different to that of smooth leather. The aim may not necessarily be to declare that one finish is better than another, but merely to note that each gives sensations that differ substantially.

The examples given are of simple products, but the concept does not change if applied to the sensations associated with complex products, such as motor vehicles. In trying out a car a user opens the car door, sits inside, touches the steering wheel and the gear change knob. Before even starting the engine the user has formulated a judgement on the vehicle, a judgement that is mainly related to tactile sensations, because the user has still not been able to try out the engine or the suspension or its comfort while traveling etc. These sensations, then, are strongly present in the relation between user and objects.

When examining a product for the first time, people unconsciously sum up the sensations that the object communicates. These are then manifest as sensations of, for example, pleasure, indifference or repulsion. Over a period of usage, people receive other stimuli from the objects and the judgement becomes more precise. Nevertheless, before being able to judge the object for its real features, there has already been made a judgement of 'pleasurableness' that heavily conditions the opinion of the product.

24.2 PLEASURABLENESS

The search for pleasure has always been at the centre of human endeavour. In the paper 'Human Factors for Pleasure Seekers' Pat Jordan wrote that: "...Since the beginning of time humans have sought pleasure. We have gained pleasure from the natural environment. From the beauty of flowers or the feeling of the sun on our skin. From bathing in soothing waters or the refreshment of a cool breeze. We have actively sought pleasure, creating activities and pastimes to stretch our mental and physical capabilities or to express our creative capabilities. Cave dwellers wrestled to test their strength and expressed themselves through painting on the walls of their dwellings. Today we 'pump iron' in the gymnasium and decorate our homes with selections of paintings and posters. Another source of pleasure has been the artefacts with which we have surrounded ourselves. For centuries humans have sought to create functional and decorative artefacts. Artefacts that have increased the quality of life and brought pleasure to the owners and users. Originally, these objects would have been clumsily bashed out from stone, bronze or iron by unskilled people who simply wanted to make something for their own use. As systems of trade and barter were developed specialist craftspeople became prevalent, creating artefacts for use by others in the community. Today, most of the artefacts that we surround ourselves with were created by industry..." (Jordan 1998).

As far back as the fourteenth century, Leon Battista Alberti wrote, in the context of architecture, that: "...furthermore, beauty is quality as such that contributes in a conspicuous manner to the comfort and even to the duration of the building, since no one can deny they feel more at ease living within ornate walls than between bare walls" (Leon Battista Alberti in Re Aedificatoria).

Recently, human factors, as a discipline, has begun to ponder ways of gauging and controlling sensations and to how they can be introduced into new products.

Many producers of consumer durables, particularly household appliances, electronic devices and above all motor vehicles, are working on quality, on the promise of supplying high levels of comfort, on the usability of the systems and the comprehensibility of the media and means of communication and learning. The devising of 'natural' controls is seen to be heading in this direction, for example with the electronic controls for motor vehicle seats, no longer formed by a long and poorly comprehensible series of round or square buttons, but taking the shape of a true and proper self-explanatory simulacrum seen to be highly effective and often even being quite beautiful to look at.

The introduction of higher quality levels in all that is perceived by the senses, all that is seen, touched or heard, as well as the undoubted sensitivity of the designer, also requires the devising and application of new and refined methods and tools. Not only should it be possible to identify pleasurable sensations for the users, but it should also be possible to rate them rigorously, propose them in projects, describe them in the related documentation, reproduce them in the objects and perfect them by comparing the proposals with reality. In the same way that it is not sensible to assume that just because the designer finds a product usable anyone can use it, the same applies to sensory aspects. It is necessary, then, to find a measure of sensorial quality that is independent from and complementary to, the judgement of the person designing the product.

At first sight this may seem to present a problem that is insurmountable, linked to the very nature of subjectivity. It is possible to examine size, weight, strain, structure — they can be measured with tools and quantified numerically with values that hold true both in time and in space; for example in considering a way of measuring weight it is possible to be certain that, within statistical limits, in another place and on another occasion the same results will be obtained (objective data), whereas in examining sensations it is not possible to obtain data with the same repeatability, because the aspects under investigation are principally subjective.

Nevertheless, all objects can be characterised by the following two types of parameter:

- objective parameters, that is the elements measurable using methods which have been elaborated by the various branches of scientific knowledge; and
- sensations of a subjective type that are closely linked to characteristics and reactions that can be gauged by techniques devised by cognitive psychology.

The results will give indications of degrees of pleasure that can be used in design. The objective is to gauge and reproduce these aspects of pleasurableness and to turn them into specifications that can be introduced into the manufacture or supply of products in the same way that the technological performances required from the materials and manufactured products are specified. This means going from the sphere of speculation to that of rigour. In other words the study of sensory quality would be approached in a manner similar to the concept of usability, so as to be able to obtain the performance promised by a product without penalising aspects of well-being or comfort and achieving pleasure in its use.

Sensory aspects

The sensory pleasure of objects concerns the qualities perceived through the senses (touch, sight, hearing, smell and kinaesthetics), that are beyond any absolute measurement that would make them valid for everyone for all time. Subjective pleasure is strongly linked to the shifting aspects of individual and cultural variables, and may thus be dependent on the culture of the ethnic and social group the subjects belong to, the sensory modalities being governed by the different individual perceptive characteristics, in turn dependent upon the elements of 'taste' of the society in general and of the individual.

The *culture* variable constituted by the social or ethnic group to which the subject belongs would mean that, for example, an Asian surely has a historical and cultural background very different to that of a European, not unlike the differences within the same grouping between someone who lives in contact with nature and a city-dweller.

The *sensory aspects* also stand as a dividing element because each individual has his or her own characteristics which derive from functional reductions that are acquired or congenital (an example would be age), or deriving from the type of work they carry out (there are surely strong differences of tactile perception between a subject who works as a bricklayer, who uses his or her hands for heavy work and another subject who uses his or her hands in a lighter way, such as a computer operator).

Questions of *time* concern the change over time of taste: for example certain colour combinations that were judged by our fathers as being extremely vulgar in that they were considered 'loud', are now seen as recherché and pleasing, not to speak of the evolution of what is appreciated as music.

Ergonomics has proposed methods and tools for analysing the subjective sensations of the sensory qualities transmitted by the object when it enters into contact with users. An example of such a technique is Kansei Engineering (Nagamachi 1995; see also Jordan, Chapter 23).

The subjective sensation felt when in contact with artefacts can be broken down into sense, sight, touch and hearing, with reference to the various characteristics of the object that we may call stimulus components:

- tactile sensations: the shape and the size of the object;

- prehensile sensations: the quality, the softness, the grippiness of the surface;
- functional sensations: the mode of use and the way it functions;
- thermal sensations: conductivity and thermal capacity;
- chromatic sensations: colour and finish of the surfaces and the chromatic nature of the object; and
- acoustic sensations: sonority of the material, action, signals and acoustic feedback.

There are also sensations of taste and smell, that are secondary for gauging the sensory characteristics of the objects. It is, however, true that a 'newness' smell can be considered more or less pleasurable according to the circumstances and can have a considerable influence on the appreciation of a product. On the other hand our approach to sensory features would be totally different if we were considering food; right from 2000 BC the Chinese attributed equal importance to three elements for the pleasure of eating: taste, smell and sight. In fact the ingredients of any dish, for example 'canton rice', are also chosen on the basis of their colour mix.

24.3 THE SENSORIAL QUALITY ASSESSMENT METHOD

A specific methodology has been developed, called the Sensorial Quality Assessment Method — SEQUAM (Bandini-Buti, Bonapace and Tarzia 1997) for rating the emotional aspects of the product, to try to gauge the sensations experienced by the users when they come into direct contact with objects. SEQUAM was first used as a method for researching and applying concepts of 'pleasureableness' in a study carried out for automobile manufacturers Fiat. The study began in 1992 and was initially applied to four types of components inside motor vehicles in immediate contact with the user, and subsequently extended to other components in the following years (Bonapace and Bandini-Buti 1997). The method enables the location and description of the parameters of the objects that affect pleasure in a group of subjects representing the users, the measuring of the aspects that constitute the parameters, a subjective rating by comparing objects of the same type with different pleasurable qualities and finally a subjective rating obtained through guided interviews.

The results are given both in quantities (percentage ratings of responses) as well as qualitatively (analysis of free responses). This is summarised in Figure 24.2.

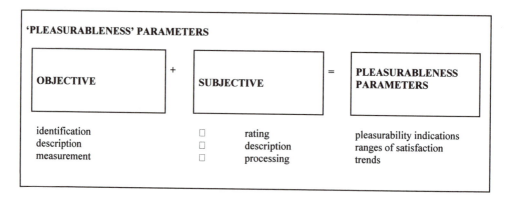

Figure 24.2 Pleasurableness parameters.

Objectives

The objectives of SEQUAM are:

- create scales showing pleasurableness that, related to the objective aspects of the product, highlight the range of satisfaction for each parameter thus leading to the formulation of indicators;
- define the pleasurableness indicators to be used by designers;
- formulate project specifications on the pleasurableness of objects and their other measurable parameters (surface, ways of gripping, softness, noisiness etc); and
- carry out predicative and correlative ratings with objective parameters to obtain transmittable data based on statistical analyses and which can be applied in the project phase.

Object of the research

The object of the research can be an object or collection of objects for a specific function (e.g. steering wheel, ambient thermostat, CD player etc.). Experience from applying SEQUAM indicates that the studies on general objects, homogenous only in terms of type of functioning (levers and switches not attached to a specific function) are to be avoided because pleasure cannot be disassociated from the mode and reason of contact. Studies on objects for the same function, though activated/used in different ways (e.g. manual and automatic gear change, driver and front passenger seats etc.), are likewise to be avoided.

The number of the objects to be tested is limited by the opportunity to organise tests with subjects. If the object is simple (e.g. a gear change knob) no more than 10 samples need be tested.

Type of trials

The Sensorial Quality Assessment Method is set out in study phases that support the gauging of the sensation of pleasure perceived by the individual. Two basically different types of trials can be carried out:

- Laboratory tests with objects isolated from their use-context and assembled on structures specifically devised for carrying out the studies. Carrying out trials studies on isolated objects in laboratories or workshops enables the testing of a relatively high number of samples, it speeds up the testing and enables optimum administration of trials times.
- Testing objects in a mock 'natural' context (household appliances in mock kitchens, parts of cars assembled on supports). These are preferable conditions in as much as they are closest to real use, although bringing the users to the trials area already creates a certain degree of artificiality (see Kanis, Chapter 4). The trials are difficult to carry out for reasons of time and costs. The condition closest to reality is the observation of the behaviour of users in their habitual place of activity. This is particularly onerous in that it is the laboratory or workshop that has to move, and often not very effective because at times only products possessed by the users can be tested.

Subjects/users

In order to obtain reliable information the subjects should reflect the characteristics of the potential users, for example, in terms of age and sex. Depending on the object of the research some parameters of choice, such as the stature and size of the individuals for research on motor vehicles and armchairs, previous experience of studies on particular topics such as professional products, the size of the hand for studies that concern the manipulation of objects etc., may also be relevant. The subjects should be chosen from people not connected with the companies that produce the product under examination and in particular they should not be testers. They should possess a certain sense of criticism and the capacity to imagine real situations starting off from more-or-less realistic simulations (ranging from the isolated object in the laboratory or workshop to the completely functioning object presented in simulated conditions of use).

Analysis of data

The data collected can be processed so as to obtain a scale of pleasurableness (from the most to the least appreciated), to process the objective data so as to obtain scales of values, to make comparisons between the subjective and the objective data for those parameters where measurements could be made (e.g. size, hardness, roughness etc) and finally to identify the topics that, due to the intrinsic limitation of the sample used, require further study. All the data gathered in the subjective and objective tests can be used in formulating project specifications or to update what was already in force. In fact the data on pleasurableness is liable to evolve in time and in space, hence the test will have to be repeated in time.

The project specifications can be represented by:

- optimum maximum and/or minimum figures;
- trends;
- figures or trends that refer to specific product or user categories (e.g. sports cars, older users);
- issues that are difficult to quantify and very much influenced by the surrounding conditions, but nonetheless have to be held in consideration and rated during the evolution of the project.

The SEQUAM methodology establishes that the work scheme should be repeated in three sequential study phases:

- research on current products;
- research on innovation;
- verifying prototypes to support the project development phase.

The SEQUAM process is summarised in Figure 24.3.

THE SEQUAM PROCESS

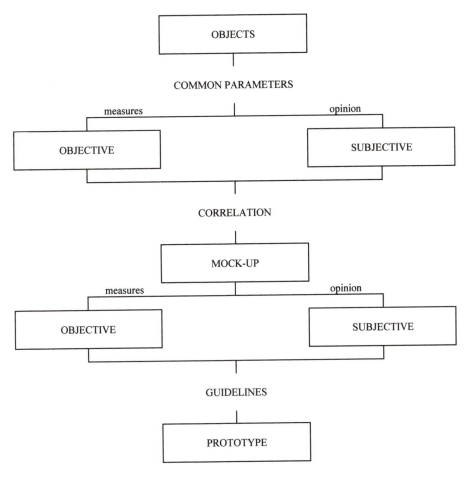

Figure 24.3 The SEQUAM process.

First phase — research on current products

The first phase constitutes a study of a selection of objects currently on the market deemed particularly interesting for their various aspects of pleasurableness — this is in order to give the broadest possible range of stimulus components. In fact the commercial output is the expression of the proposals and the creative verve of numerous designers and technicians. The purpose of this phase is to quantitatively rate the appreciation of the stimulus components and to direct the design of the experimental study models needed for a systematic study of the pleasurableness of components.

The research on current products is set out as follows:

1. Definition of the sensory parameters that characterise the object under examination (e.g. touching, rotating, opening, closing etc) and identification of the objective qualities that can be measured and manipulated (e.g. size, weight, surface finish etc). This study is carried out by the researchers themselves and by a small group of users in direct contact with the object, in order to identify the parameters to be taken into consideration as well as the priorities.

2. Selection of objects already produced that in various degrees possess the qualities being studied and a choice of a certain number (from 5 to 10 depending on the given complexity and the duration of the study with the subjects) that is meaningful for the parameters that are being studied. Each object is to be defined by its measurable (size, hardness, weight, sound, excursion etc) or qualitative features (family of shapes, sameness of shapes, surface characteristics, colours, features of action, noise features etc.).

3. Choice of subjects representative of the final user and if necessary of any distinctive characteristics of the human/object relationship (e.g. size of the hand). Selection of at least 18 subjects that can be grouped in terms of sex and age. The current trend is to work with an even greater number of subjects. The subjects should not be specialists with respect to the object being studied, such as testers, service personnel, employees of the producer company, because they could be conditioned by familiarity with the object or by private interests. The subject cannot generally be seen to represent the user, because it is necessary to have a dialogue with individuals capable of simulating the experience of study use conditions that are different to those of actual working conditions.

4. Carrying out the study in real life conditions or in a laboratory or workshop where conditions as close as possible to the real conditions of use have been reconstructed, showing the objects/allowing the objects to be tried out in arbitrary sequence for each question, using research protocols for rating the comparisons between the object parameters. The tests include:

- Observation, contact, the use of the object with or without the request for some special action depending on the nature of the object and the purpose of the research. The researcher observes, takes notes or records.
- Filling in a semi-structured questionnaire on free topics, on priorities and on the most important features.
- Debriefing interview to glean free impressions.
- Carrying out laboratory analysis to numerically characterise the parameters of what was studied in the subjective analyses. The analysis of some parameters can be highly specialised (surface features, analysis of noise etc).
- Processing of data obtained. Organising them in diagrams in which the subjective data (scale of pleasurableness) are correlated to the objective data (measure of the pleasurableness degree parameters of the objects). Each parameter will have its diagram with the pleasurableness ratings placed vertically and the various objects tested placed horizontally in order of pleasurableness or in other ways that are deemed important. The diagrams should always be confirmed by and compared with the impressions and suggestions gathered and with the filmed sequences. Figure 24.4 gives examples of the relationships between pleasureableness and particular product properties.

The advantages and disadvantages of analysis of current products are summarised in Table 24.1.

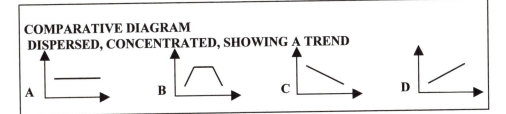

Figure 24.4 Here are examples of the types of relationship that can be observed, between an object's objective properties and the level of sensorial pleasure associated with the object. The x-axes represent the objective property (e.g. size, weight, surface finish etc.), while the y-axes indicate the level of associated sensorial pleasure. The diagrams can basically show a broad dispersion (A: The parameter does not appear to be essential in terms of size), a central peak (B: The parameter is not important in determining the pleasurableness), an off-centre peak (C/D: A trend emerges where the inversion of the trend cannot be read).

Table 24.1 Study on manufactured products: advantages and disadvantages.

ADVANTAGES
- Samples are readily available for testing.
- All the variables can be tested because the object is finite and functioning.

DISADVANTAGES
- The object may conjure up stereotypes in the interviewee that are difficult to eliminate (well-known or famous objects that the subject knows better than the others or that he/she uses habitually etc).
- Only usual and non-innovatory parameters can be tested.

The first phase of the study gives the first data on the 'pleasurableness' of the type dealt with, allows confirmation of ideas about the sensory categories and enables identification of the parameters that can be gleaned with greater precision.

Second phase — study on innovation

The study of what is on the market gives indications as to the pleasurableness of what is currently being produced and is thus meaningful for the here and now; however it does not tell us anything about innovatory trends. Above all, it should be noted that what is offered on the market today has been conceived a fair time back (even up to five years ago for motor vehicles). It is hence necessary to introduce a study phase where objects purposefully designed for research are analysed. This must enable:

- the research of parameters that in the first phase required further going into;
- the analysis of objects that completely respond to the preferences that emerged in the preceding phase;
- the analysis of the trends that emerge from the the first part of the study but were not followed through due to a lack of suitable objects; and
- the study of innovatory trends.

This phase consists of planning, creating and experimenting with a homogeneous series of study models and in reading the responses to the varying of individual object properties in 'low noise' situations; that is, in situations not polluted by preconceptions. This phase is needed to check the variability limits of the components studied by submitting models that may have exaggerated features for judgement (for example grips that are exceptionally large or exceptionally small).

The method of study is the same as that applied to the first phase and can be carried out using the same subjects. The questions are the same as those posed in the first phase but with the addition of the developments programmed in the design phase and with adjustments required by the technical features of the models.

Table 24.2 summarises the advantages and disadvantages of the analysis of study models.

Table 24.2 Research on study models: advantages and disadvantages.

ADVANTAGES
- With the creation of specially devised models the single parameters can be analysed to great effect by comparing the models of objects that are identical in all their parts apart from the parameter under study (for example the cross-section of the grip if the object is to be carried or transported).
- The time required for developing this phase is not long even if slightly greater than that of the preceding phase in that projects need to be developed, and models in wood and sometimes other materials need to be constructed.

DISADVANTAGES
- The costs and time for carrying out the proceedings mean that study projects can generally only be made on non-functioning models.

Some parameters, such as acoustic and thermal characteristics and the functioning of mechanisms require the application of the same method but with specific equipment which in some cases is fairly sophisticated.

The results of these two phases establish a series of project guidelines on the pleasurableness of the types studied that are valid for the here and now, hence to be checked in time as consumer expectations and general features evolve.

It is important to note that the projects devised during the study should not be seen as true and proper design projects in as much as they have been conceived for study purposes (hence for checking and very often for exceeding the limits of pleasurableness) and not as part of an organic design context. The results of the study are to be seen as a stimulus and as guidelines for the overall design project, and as a means for rigorously gauging the aspects of pleasurableness, something that, when entrusted to the sensitivity of the single person as is generally the case, is not easy to transmit.

The figures obtained are then divulged by the devising of project guidelines that also bear a detailed account of the criteria for inserting 'pleasurableness' in the design.

Third phase — verifying the prototype

In the first two phases, the study is carried out in conditions that should allow the researcher to get the maximum information with limited development times, as is generally required by manufacturing industry. These conditions are to be defined as 'workshop' or 'laboratory' conditions because even in the most favourable instances where perfectly functioning products are used, the studies are carried out in conditions that are different from the real conditions of use. Under the experimental conditions the subjects concentrate on the objects as such, used in the presence of an observer, while in actual usage the subjects are only aware of the results of their interactions with the products.

For the final check of the pleasurable features of the products, one has to be able to carry out studies on functioning prototypes that have all of the features of the finished products and that are inserted in coherent surroundings from both a functional and a formal point of view, and used in actual conditions by a cross-section of potential users.

When the development of the projects allows the production of prototypes that are fairly close in terms of finish and function to the finished product, the results of the first two phases can be checked in real use conditions by a larger cross-section of subjects.

During the project development of the product, which should include the ergonomists as part of the design group, the various stages of the product can be checked, combining the tests with the data collected in the two phases.

24.4 COMMUNICATION OF ERGONOMIC STUDY RESULTS IN LARGE INDUSTRIES

The industries that produce on a large scale feature an ever more complex system of organisation, entailing huge investments, and are subject to pressing demands from the market. Thus it is paramount that the concept of ergonomics and pleasure and the results of studies enter into normal procedures, so that they can be introduced into the development of products offering the slimmest of uncertainty margins.

The concepts of pleasurableness are effective if they are communicated to planners and designers as parameters that are part of a systematic and transmittable approach. The objective 'pleasurableness' guidelines are proposed to the product development team as elements that qualify the project and as tools for introducing pleasure into their projects, with guidelines that do not limit the creative freedom. Testing on prototypes enables the effectiveness of the guidelines to be checked.

Another element of great importance for mass-production industries is that the application of the method supports dialogue between suppliers, product managers and designers on sensory quality which is based on objective data.

Targeted and more generalised studies

Internal industry studies for introducing ergonomics into projects can be targeted to a specific product under development or cover all present and future ones. In both cases it will be necessary to start up research and study activity on usability and pleasure involving a certain number of subjects, activity that requires complex methods and a suitable amount of time and resources for its completion.

When the development of a specific model is the objective, a work group is formed in which close contact is established between the ergonomic researchers and the persons directing the project and the formal development of the model, as well as with marketing services, suppliers and buying offices. This makes communication of the study results easier even though these may not be expressed in a strictly formal way.

The formal nature becomes indispensable when information from present and future products is drawn on in the study — here the communication of the results of ergonomic studies, principles and techniques has to be rapid and direct at all company levels affected by or involved in the project. It is important to stimulate things in the direction of the ergonomics of structures that make up the complex system of design, enabling access in real time to know-how that has to be simple to grasp, in as much as few of the potential users have had direct experience with basic ergonomic research. The reports offering the conclusions of the study, full of information, descriptions and working details, are often difficult to read and at a time when use of this data is, hopefully, becoming a daily feature of all the stages involved in developing and testing a product.

Computerised documentation

The final reports of the study carried out with SEQUAM methodology are in the form of complex printable documents, containing all of the elements needed to tackle the subjects and to transmit the working conclusions. However, growing demand for ergonomic characterisation of the product has led the organisers to represent the results in a simpler, more communicative form that can be accessed and used at different levels. This has led to:

- the transmission of data in a more synthetic, more easily readable manner for users with different backgrounds (technicians, designers, stylists and administration);
- the use of rapidly consultable charts and diagrams; and
- readability at different levels, enabling easy reference to the background study documentation and access to study data that justify the results obtained.

The properties of effective information tools for use in this context are summarised in Table 24.3.

Table **24.3** Usability and pleasurableness studies: information tools.

Product ergonomics

➔ **EFFECTIVE TRANSMISSION** (for technicians stylists, administration)

➔ **RAPID CONSULTATION** (graphic form)

➔ **READING ON A NUMBER OF LEVELS** (parameters, descriptions, guidelines)

Figure 24.5 gives an example of how the output from a SEQUAM study could be presented.

The study on the aspect of pleasurableness gleaned in the various studies made up to now in the field of car interiors have been collected together in a tree information system being listed under subjects:

PARAMETERS

Listing of the parameters relative to the subject, that is those aspects that from an ergonomic point of view characterise the object or the system. Degree of importance attributed to them by the subjects during testing is also defined.

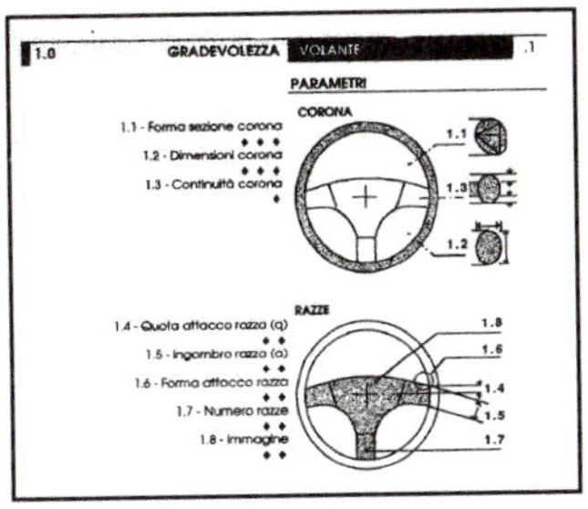

CATEGORIES

Conceptual categories (formal, use and sensorial) to which all similar systems and objects can be referred to.

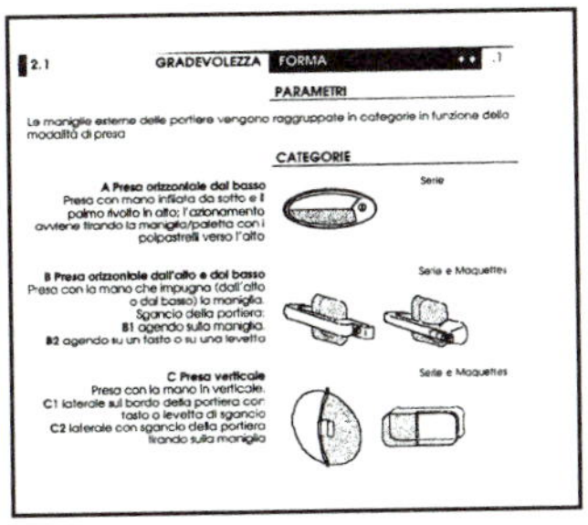

DESCRIPTION OF TYPES

Representation in diagram form for the comparison of various types analysed (description of types in SAE grading).

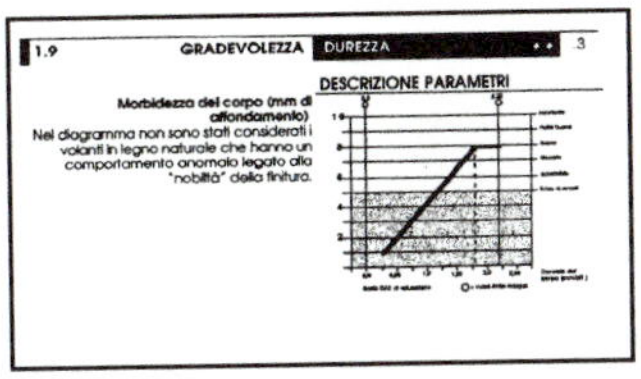

DESCRIPTION OF PARAMETERS

Representation in diagram form of the various parameters examined (description of parameters in SAE grading).

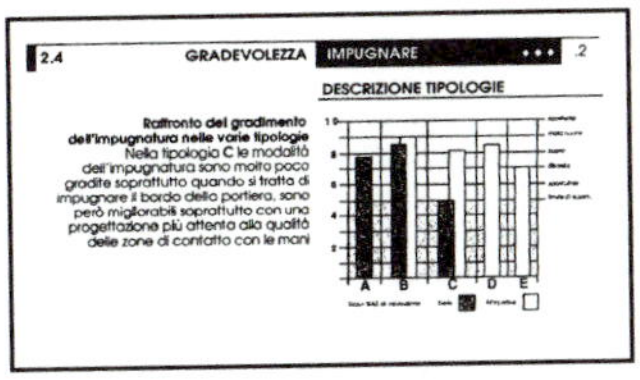

GUIDELINES

Pleasurableness guidelines.

The documents have been configured so as to make them available to all the bodies involved in the projects on the companies internal computer network.

Figure 24.5 Images from the Bologna report on the intranet document, computer documents.

24.5 REFERENCES

Bandini-Buti, L., Bonapace, L. and Tarzia, A., 1997. Sensorial Quality Assessment: a method to incorporate perceived user sensations in product design. Applications in the field of automobiles. In *Proceedings of IEA 1997*, Finnish Institute of Occupational Health, Helsinki, pp. 186-189.

Bonapace, L. and Bandini-Buti, L., 1997. The pleasant object: the Sensorial Quality Assessment Method. *Ergonomia*, **10**, Moretti & Vitali, Bergamo, pp. 4-14.

Jordan, P.W., 1998. Human factors for pleasure seekers. In *Ergonomia*, **11**, Moretti & Vitali, Bergamo, pp. 14-19.

Nagamachi, M., 1995. *The Story of Kansei Engineering*. Kaibundo Publishing, Tokyo.

CHAPTER TWENTY-FIVE

Discussion and Conclusion

BILL GREEN

Faculty of Industrial Design Engineering, Delft University of Technology,

Jaffalaan 9 2628 BX Delft, The Netherlands

and

PAT JORDAN

Philips Corporate Design, Building W, Damsterdiep 267,

P.O. Box 225, 9700 AE Groningen, The Netherlands

25.1 INTRODUCTION

The discipline of ergonomics/human factors is at an important point in its development. It is a discipline which has always been linked with design, whether of work practices or products or systems, but developments in the recent past have placed a new emphasis on the nature of those links. Domestic and other products have always to some extent been the subject of anthropometric and biomechanical scrutiny, at however crude a level, and other aspects of human characteristics have been routinely employed as limiters in the design of complex systems, usually for expert or trainable users. The electronic revolution has precipitated an unprecedented invasion of hitherto esoteric problems into the common domain, and together with the greying of the population of that domain, has created new demands on ergonomists and the design profession. So-called 'intelligent products' have been seen sometimes to be remarkably stupid, and the strategies of traditional ergonomics revealed as inadequate to deal with the problem. The statement that "allowing the possibility of use says nothing about how a product will be used" (Green, Chapter 10) may be seen as a truism, but it helps to focus on the changes which are taking place in ergonomics and design research and practice. The user as a set of quantifiable characteristics is giving way to the user as an unpredictable part of a complex interaction, and in turn the emphasis is being placed on researching what has been termed the ecology of that interaction. This ecology is never more complex than when domestic products and users are the central issue. The view of ergonomics taken in this book is a broad one, and steps over the boundaries of the classical work efficiency definitions to deal with the complete user/product interaction. An attempt has been made to give an overview of some of the best of current practice and to identify a number of issues that the profession will be called upon to address in the years to come. Whilst the discipline is currently thriving to an unprecedented degree (Jordan, Chapter 21), its continued success is likely to be dependent on the effectiveness with which it meets these new challenges.

From the contributions in this book, there emerge some major issues that the profession must address if it is to continue to flourish.

25.2 THAT WAS THEN, THIS IS NOW!

From the point of view of customers, usability has moved from being a 'satisfier' to a 'dissatisfier'. A few years ago, users would probably have accepted low levels of usability, provided that a product performed well, contained useful functions, was durable and so on. If a product had good levels of usability, users would have seen this as a plus point (in marketing terms, as a 'satisfier'). The increasing emphasis in the popular media on what started out being called 'user friendliness' — usually applied to computers only but now a point of critical judgement of almost anything — has led to the buying public tending to take it for granted that a product will be usable and 'user friendly', and being dissappointed if it is not (in marketing terms a 'dissatisfier'). The concept of 'friendliness' is flexible, and Jordan (Chapter 21) argued that if human factors is to help create positive benefits that users will really notice, it must now move beyond usability, to more holistic 'pleasure-based' approaches. Bonapace (Chapter 24) and Taylor (Chapter 22) gave examples of such approaches and their practical application, and Jordan (Chapter 23) reviewed the Japanese technique of Kansei Engineering, a technique for analysing the relationship between a product's design properties and users' emotional reactions to the product.

25.3 GETTING OLDER

A second major issue is the aging of the population. The elderly are becoming an increasingly large and financially powerful group. Many years ago, Victor Papanek, in his pioneering work 'Design for the Real World', made the point that a collection of minorities with a common problem quickly become a majority. We are now witnessing a single commercial minority growing into a powerful force. As Coleman (Chapter 16) noted, 35% of UK consumer spending is currently controlled by those over 50 years of age. It is clear from the work reported by Etchell (Chapter 19), that many manufacturers are still putting products onto the market with little or no thought for the requirements of their elderly customers. Coleman has outlined a number of strategies for approching design for an aging population and Freudenthal (Chapter 20) laid down a few basic guidelines for designing for the elderly. Jordan (Chapter 17) demonstrated that design for the elderly can be included as part of an inclusive approach to design, which supports the creation of products that can be used by all, regardless (in as far as possible) of the user's age or (dis) abilities. It was argued that inclusive design approaches were both a financial and a moral imperative for the profession. There is no doubt that the issues are complex, and that there is no glib and comprehensive formula for dealing with them, but it is equally clear that use of the methodologies of user trialling and usability testing promulgated in the previous chapters can assist designers to make decisions which are sustainable from both a sociological and a commercial point of view.

25.4 USING THE RIGHT TOOLS IN THE RIGHT PLACE

The third major issue was the role of human factors in the product creation process. Human factors techniques must be relevant and practical in a commercial manufacturing context. Green (Chapter 1) described some of the conditions for the successful integration of usability issues in product development. In particular, warning against over reliance on exhaustive 'experimental' trialling — an approach which can prove extremely costly. Popovic (Chapter 3) and Sade (Chapter 7) gave an overview of a range of different evaluation techniques, explaining their strengths and weaknesses and the situations under which each is best used. It is surely important that those carrying out usability evaluations have a portfolio of techniques so that they can approach each project in the most appropriate and cost effective manner. Demonstration of appropriate techniques has been focussed on in case studies in this book, for the simple reason that theory development is lagging behind the empirical process. Green (Chapter 10), Kanis (Chapter 4) and Rooden (Chapter 14) have all raised the issue of the lack of a substantial theoretical approach to user/product interaction. This lack has the effect of slowing down progress to the pace at which a new user trial becomes absorbed into the usability lexicon and contributes a new insight. Green and Kanis (1998) have explored this issue in another forum and we can only support their conclusion that attempting to add to the 'body of findings' is the best contribution anyone can make. To borrow a phrase from another discipline, we must 'think globally and act locally'.

An issue that follows the academic search for a sound theoretical base is the tension between academe and industry, and this emerged very cleary during the discussion at the Tampere conference. Such tensions are, of course, not new, and are part of the fabric of most disciplines. However, domestic product development is more vulnerable to such tensions than some, as a consequence of the actual closeness of the fields. The connections between academic research and industrial practice in usability are so evident that when they don't work well, it is plain to all involved. In Tampere, the divergence between the rigour of academe and the pragmatism of industry was debated by a group of remarkably homogenous people: ergonomists whose primary interest was design, and designers with a strong background in ergonomics. At one point it was described as a 'war' and this reflected much of the discussion. The war was no less bloody for being a civil one! The positions may be summed up as follows: academics regard industrial approaches as sloppy and lacking in rigour and validity, while industrialists regard academic practice as over complex and impractical. They want 'off the shelf' methods which are instantly applicable and have some acceptable level of validity. A common complaint was that academics have been slow to respond to these requirements, and that the tools and techniques are 'sledgehammers to crack nuts'.

25.5 WORKING TOGETHER

Finally, it is worth re-stating the obvious: The relationship between the designer and the human factors specialist is central to the creation of products which can fulfill usability criteria in their broadest sense. Thomas and van Leeuwen (Chapter 11) demonstrate how close co-operation between interaction designers and human factors specialists can lead to the creation of a product widely recognised as having achieved the very highest standards of usability. This relationship operates best at an individual level when a designer, design engineer or product developer has a strong background in ergonomics and experience of trialling techniques (it seems, unfortunately, rare for a human factors specialist to have a strong design background) and, at a team level, when the design and development efforts are conducted in a mutually supportive and non-hierarchical environment, with a management who recognises the important competitive advantage to be gained from truly excellent design for people.

25.6 REFERENCES

Green, W.S. and Kanis, H., 1998. Product interaction theory. In *Global Ergonomics*, P.A. Scott, R.S. Bridge and J. Chartriss (Eds), Elsevier, Amsterdam, pp. 801-806.

Index